# STELLAR EVOLUTION AND NUCLEOSYNTHESIS

# Documents on Modern Physics

Edited by

ELLIOTT W. MONTROLL, University of Rochester
GEORGE H. VINEYARD, Brookhaven National Laboratory
MAURICE LÉVY, Université de Paris

---

A. H. COTTRELL, Theory of Crystal Dislocations
A. ABRAGAM, L'Effet Mössbauer
A. B. PIPPARD, The Dynamics of Conduction Electrons
K. G. BUDDEN, Lectures on Magnetoionic Theory
S. CHAPMAN, Solar Plasma, Geomagnetism and Aurora
R. H. DICKE, The Theoretical Significance of Experimental Relativity
BRYCE S. DEWITT, Dynamical Theory of Groups and Fields
J. W. CHAMBERLAIN, Motion of Charged Particles in the Earth's Magnetic Field
JOHN G. KIRKWOOD, Selected Topics in Statistical Mechanics
JOHN G. KIRKWOOD, Macromolecules
JOHN G. KIRKWOOD, Theory of Liquids
JOHN G. KIRKWOOD, Theory of Solutions
JOHN G. KIRKWOOD, Proteins
JOHN G. KIRKWOOD, Quantum Statistics and Cooperative Phenomena
JOHN G. KIRKWOOD, Shock and Detonation Waves
JOHN G. KIRKWOOD, Dielectrics—Intermolecular Forces—Optical Rotation
HONG-YEE CHIU, Neutrino Astrophysics
M. TINKHAM, Superconductivity
DAVID HESTENES, Space-Time Algebra
J. A. WHEELER, Geometrodynamics and the Issue of the Final State
R. H. DALITZ, The Quark Model for Elementary Particles
P. H. E. MEIJER, Quantum Statistical Mechanics
H. REEVES, Stellar Evolution and Nucleosynthesis
J. LEQUEUX, Physics and Evolution of Galaxies
V. KOURGANOFF, Introduction to the General Theory of Particle Transfer
L. SCHWARTZ, Application of Distributions to the Theory of Elementary Particles in Quantum Mechanics

*Additional volumes in preparation*

# Stellar Evolution
# and Nucleosynthesis

HUBERT REEVES

Professeur à l'Université de Montréal
Détaché à l'Institut d'Astrophysique de Paris
et à l'Institut de Physique Nucléaire d'Orsay

GORDON AND BREACH
*Science Publishers*
NEW YORK · LONDON · PARIS

Copyright © 1968 by Gordon and Breach, Science Publishers Inc.
150 Fifth Avenue, New York, N.Y. 10011

Library of Congress Catalog Card Number: 67-28239

*Editorial Office for the United Kingdom:*
Gordon and Breach, Science Publishers Ltd
8 Bloomsbury Way
London W.C.1

*Editorial Office for France:*
Gordon & Breach
7–9 rue Emile Dubois
Paris 14e

Distributed in France by:
Dunod Editeur
92 rue Bonaparte
**Paris 6e**

Distributed in Canada by:
The Ryerson Press
299 Queen Street West
Toronto 2B, Ontario

Printed in Great Britain by Butler & Tanner Ltd, Frome and London

## Editors' Preface

SEVENTY years ago when the fraternity of physicists was smaller than the audience at a weekly physics colloquium in a major university a J. Willard Gibbs could, after ten years of thought, summarize his ideas on a subject in a few monumental papers or in a classic treatise. His competition did not intimidate him into a muddled correspondence with his favourite editor, nor did it occur to his colleagues that their own progress was retarded by his leisurely publication schedule.

Today the dramatic phase of a new branch of physics spans less than a decade and subsides before the definitive treatise is published. Moreover modern physics is an extremely interconnected discipline and the busy practitioner of one of its branches must be kept aware of breakthroughs in other areas. An expository literature which is clear and timely is needed to relieve him of the burden of wading through tentative and hastily written papers scattered in many journals.

To this end we have undertaken the editing of a new series, entitled *Documents on Modern Physics*, which will make available selected reviews, lecture notes, conference proceedings, and important collections of papers in branches of physics of special current interest. Complete coverage of a field will not be a primary aim. Rather, we will emphasize readability, speed of publication, and importance to students and research workers. The books will appear in low-cost paper-covered editions, as well as in cloth covers. The scope will be broad, the style informal.

From time to time, older branches of physics come alive again, and forgotten writings acquire relevance to recent developments. We expect to make a number of such works available by including them in this series along with new works.

ELLIOTT MONTROLL
GEORGE H. VINEYARD
MAURICE LÉVY

# Stellar Evolution and Nucleosynthesis

A SERIES of lectures given during the fall of 1964, at the Fondation Universitaire, 11 rue d'Egmont, Brussels, by Hubert Reeves, of the Department of Physics of the Université de Montréal and of the Goddard Institute for Space Studies, N.A.S.A., New York City, under the auspices of the Institut Interuniversitaire des Sciences Nucléaires and of the Organisation for Economic Co-operation and Development.

The same lectures were given in the Spring of 1965, at the Institut d'Astrophysique of the Université de Paris, 98 bis Boulevard Arago, Paris XIV$^e$.

# Acknowledgments

MESSRS. Raymond Coutrez and Marcel Demeur made possible my stay in Europe. Mr. Marcel Demeur organized the series of lectures given in Brussels. Mr. Evry Schatzman organized the Paris series. Miss Monique Crétin, Messrs. Guy Reidemeister and Robert Van den Borre, of the Université Libre de Bruxelles, and Mrs. Françoise Fréhel of the Institut d'Astrophysique helped in writing up the notes. Mrs. Marie-Thérèse Duchêne patiently typed them and the I.I.S.N. edited them. I am very grateful to each of these persons.

HUBERT REEVES

THE stay of Professor Hubert Reeves and the organization of the series of lectures, the text of which forms the present volume, were made possible by the financial assistance of the Institut Interuniversitaire des Sciences Nucléaires and the Organisation for Economic Co-operation and Development.

We wish to express our deep gratitude to the Members of the Commission Scientifique of the I.I.S.N. and also to Mr. Deloz, Conseiller au Ministère, for the help they gave us.

RAYMOND COUTREZ<br>MARCEL DEMEUR

## Table of Contents

|  |  | Page |
|---|---|---|
| | Editors' Preface | v |
| | Publisher's Note | vii |
| | Acknowledgments | viii |
| | Introduction | xiii |

| | | |
|---|---|---|
| *Lecture I* | The role of the four fundamental interactions of physics in stellar and galactic evolution | 1 |
| | I–A: The initial state: energy balance | 1 |
| | I–B: The gravitational interaction; the hydrostatic equilibrium states | 3 |
| | I–C: The electromagnetic interaction: contraction and luminosity | 4 |
| | I–D: Evolution in the Pressure–density plane; limiting values | 6 |
| | I–E: Nuclear and weak interactions | 8 |
| |     I–E–1: Energy production | 8 |
| |     I–E–2: Neutrino emission | 9 |
| |     I–E–3: Nucleosynthesis | 9 |
| | I–F: The stages of thermonuclear fusion | 10 |
| | I–G: Electromagnetic interaction coupled with weak interaction: pair formation; neutrinos | 12 |
| | I–H: Electromagnetic interaction: nuclear photodisintegration | 14 |
| | I–I: Final energy balance | 17 |
| | I–J: Galactic evolution; nucleosynthesis; stellar populations | 17 |
| | Bibliography | 20 |
| *Lecture II* | The role of charged particles | 22 |
| | II–A: The mapping between the $Z$–$N$ plane and the H–R plane | 22 |

|  | Page |
|---|---|
| II–B: Nuclear reactions: generalities | 23 |
| II–C: Nuclear reactions: charged particles | 24 |
|     II–C–1: Non-resonant rate | 25 |
|     II–C–2: Resonant rate | 27 |
|     II–C–3: Multiresonant reactions | 30 |
| II–D: Photodisintegration rates | 31 |
| II–E: Nuclear reactions in equilibrium with the photon gas | 33 |
| II–F: The electronic screen | 34 |
| II–G: Stellar models | 36 |
| II–H: The Hertzsprung–Russell diagram | 37 |
| II–I: The H–R paths | 39 |
| II–J: Duration of the "Main Sequence" phase | 42 |
| II–K: The mass–luminosity relation | 42 |
| II–L: The red-giant phase | 45 |
| II–M: Shell sources | 45 |
| II–N: Flash | 46 |
| II–O: The helium burning | 46 |
| II P: New contraction phases in neutrinos | 50 |
| II–Q: The carbon burning | 51 |
| II–R: Neon photodisintegration | 53 |
| II–S: The oxygen burning | 53 |
| II–T: The mapping effect | 53 |
| Bibliography | 56 |
| *Lecture III* The role of gammas and neutrinos; the iron peak | 58 |
| III–A: Neutrinos: the shadow of a doubt | 58 |
| III–B: Neutrinos: simplified calculation | 59 |
|     III–B–1: Photoneutrinos | 60 |
|     III–B–2: The pair annihilation mechanism | 61 |
|     III–B–3: Plasma oscillations | 61 |
| III–C: Neutrinos and the H–R paths | 62 |
| III–D: Photodisintegration: equilibrium ($e$) process: the iron peak | 65 |
| III–E: The time scale of the $e$ process | 67 |
| III–F: Neutrinos and the iron peak | 68 |
| III–G: The end of the sequences of states of gravitational equilibrium | 69 |

TABLE OF CONTENTS        xi

|  | *Page* |
|---|---|
| III–H: Supernovae (S.N.) | 70 |
| Bibliography | 71 |

*Lecture IV* The role of neutrons: heavy elements synthesis    72
- IV–A: Why neutrons?    72
- IV–B: Irradiation parameters    72
- IV–C: Neutron sources in stars    75
  - IV–C–1: Before the hydrogen-burning phase    75
  - IV–C–2: The contraction phase preceding the helium burning    76
  - IV–C–3: Helium burning    77
  - IV–C–4: Carbon burning    78
  - IV–C–5: Oxygen burning    78
  - IV–C–6: Later sources    79
  - IV–C–7: Source characteristics    79
- IV–D: Heavy elements    79
- IV–E: The $s$ elements    82
- IV–F: The $r$ elements    86
- IV–G: Location of the $s$ and $r$ mechanisms    88
- IV–H: The $p$ elements    88

Bibliography    89

*Lecture V* The evolution of the galaxy; stellar statistics and cosmochronology    91
- V–A: Definitions and theoretical programme    91
- V–B: Simple model; the density function of the main sequence    92
- V–C: Rate of star creation    93
- V–D: Evolution of the galactic gas    94
- V–E: Nucleosynthesis of the groups of elements    94
- V–F: Nucleosynthesis of the unstable isotopes    96
- V–G: The age of the $r$ elements and the age of the galaxy    96
  - V–G–1: Hypothesis of sudden synthesis    97
  - V–G–2: Hypothesis of continuous and constant synthesis    97
  - V–G–3: Hypothesis of continuous synthesis connected with the galactic activity    98

Bibliography    99

# Introduction

IN astrophysics we assume that the laws of physics, obtained in general by studying local (atomic, nuclear, etc.) phenomena, are applicable, by extrapolation, to the whole universe. With this hypothesis, we proceed as far as possible in the analysis of the "global behaviour" of the universe.

In this "global behaviour", two related phenomena seem to play a very important role: stellar evolution and nucleosynthesis. The observable universe is neatly divided into a great number of galaxies, and each galaxy contains a great number of stars. These stars are born, live and die. A mass of gas becomes isolated, condensed onto itself, irreversibly transformed by the action of the interactions of physics, then returned (at least partially) to the galactic interstellar gas. The cycle starts again, and each cycle leaves its mark on the aspect of the galaxy. The galaxy becomes older.

In these lectures, we wish to study these two phenomena in detail, and try to determine the individual role of each of the interactions of physics, then the role of various groups of elementary particles in stellar, nucleosynthetic and galactic evolution.

The next step would be the study of the role of galactic evolution in the evolution of the universe. We know almost nothing about this question. It is the new field of study which astrophysics is already preparing to tackle.

# I

## The role of the four fundamental interactions of physics in stellar and galactic evolution

### I–A: *The initial state: energy balance*

WE first consider a problem which is fictitious, but close enough to reality to teach us many things. We consider a mass of $10^{55}$ to $10^{57}$ hydrogen atoms (almost all stars have masses lying within these limits and, at birth, the amount of hydrogen atoms generally exceeds 90%). In its initial state our "star" is cold ($T = 0$), it has infinite radius, it is neutral and is the seat of no magnetic field. Its energy balance is as follows:

a) Since the radius is infinite, we will make the convention that the total gravitational energy is zero: $\Omega_g = 0$.
b) The total kinetic energy is zero: $T = 0$; $E = 0$.
c) The nuclear potential energy is given by the sum of the masses of the proton and of the electron: $\Omega_N = 938\cdot3$ MeV/nucleon (in order to facilitate bookkeeping, energy will always be given "per nucleon").

The four interactions of physics are described in the table; the second column describes the particle which carries the interaction, the third describes its intensity, the fourth describes its range in fermis ($1\,f = 10^{-13}$ cm) and the fifth gives an example of its activity.

Because of their short range, the nuclear and weak interactions are not active in a diluted mass. Since the gas is neutral and there are no magnetic fields, the electromagnetic interaction (E.M.I.) is also not active. Only the gravitational interaction (G.I.) comes into play. The mass contracts. In order to illustrate the role of each interaction during stellar evolution, we will arbitrarily and fictitiously keep the other three interactions "switched off" as long as possible and "switch them on" only when necessary.

Fig. I–1. Young stars surrounded by nebulosity.

| Interaction | Vector | Intensity | Range | Example |
|---|---|---|---|---|
| N.I. (nuclear) | $\pi$ meson | $\simeq 10$ | short $\simeq 2$ f (1 f = $10^{-13}$ cm) | Nuclear reactions |
| E.M.I. (electro-magnetic) | photon | 1/137 | long $1/r^2$ | Electrical repulsion $\gamma$ Radioactivity |
| F.I. (Fermi or weak) | ? | $\simeq 10^{-13}$ | short $\simeq 2$ f | $\beta$ Radioactivity Neutrinos |
| G.I. (gravitational) | ? | $\simeq 10^{-39}$ | long $1/r^2$ | Stellar contraction |

I-B: *The gravitational interaction; the hydrostatic equilibrium states*

We suppose that the stellar mass has contracted from $R = \infty$ to $R$ by a series of hydrodynamic equilibrium states. The characteristics of these states can be studied by means of the virial theorem. The following equation expresses the hydrostatic equilibrium:

$$\frac{dP}{dr} = \frac{-\rho G M_r}{r^2} \qquad \text{I-1}$$

$P$ is the pressure, $\rho$ the density, $G$ the gravitational constant and $M_r \equiv \int_0^r \rho 4\pi r^2 \, dr$. Equation I-1, multiplied by $\int_0^R 4\pi r^3 \, dr$ and integrated by parts, gives:

$$3 \int_0^R P \, dV = \int_0^R \frac{\rho G M_r \, dV}{r} = -\Omega_g(R) = \Delta\Omega_g(R) \qquad \text{I-2}$$

Now the pressure $P$ is always proportional to the kinetic energy density $u$: for a non-relativistic gas $P = \frac{2}{3}u$. (In particular, for a non-relativistic, non-degenerate, monoatomic gas:

$$P = \frac{\rho k T}{\mu} \quad \text{and} \quad u = \frac{\frac{3}{2}\rho k T}{\mu}$$

where $\mu$ is the number of nucleons per gas particle (including the electrons); $\mu = A/(A + Z)$; for hydrogen $\mu = \frac{1}{2}$.) For a relativistic gas $P = \frac{1}{3}u$.

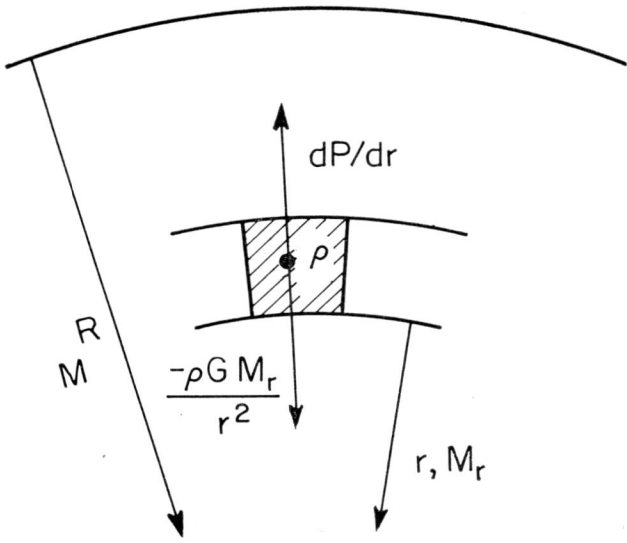

In the beginning of the contraction our gas is non-degenerate and non-relativistic:

$$+\Delta\Omega_g(R) = 2\int_0^R u\,dV = 2E_{th}(R) \qquad \text{I-3}$$

In other words, half the emitted gravitational energy has been transformed into kinetic (thermal) energy. What happened to the other half? It must have left the star. Conclusion: Contraction by a series of states in thermodynamic equilibrium requires the emission of half of the released energy! We need a mechanism for energy emission.

### I–C: *The electromagnetic interaction: contraction and luminosity*

We "switch on" the electric charges (E.M.I.) of our particles. The mass contraction leads to the acceleration of the charged particles, which leads to the production of photons. In order to be "emitted" by the star, the photons must reach the surface. On their way the photons are absorbed and re-emitted many times. Since the contraction rate depends on the emission rate, it is clear that the phenomena of radiation transfer and the opacity property ($\kappa$) of the stellar medium will play a primordial role in stellar evolution.

The production of photon energy is expressed by the first law of Thermodynamics (equivalent to $dQ = -dU - P\,dV$).

$$\varepsilon_{ph} = \frac{P}{\rho^2}\frac{d\rho}{dt} - \frac{du}{dt} \qquad \text{I-4}$$

$\varepsilon_{ph}$ is the rate of production of photons per gm/sec. The second term represents the work done by the contraction and the third is the internal energy gradient.

The rate of radiation transfer is determined by the relation

$$\frac{L_r}{4\pi r^2} = \frac{-1}{3\kappa\rho}\frac{d}{dr}(acT^4) \qquad \text{I-5}$$

where $L_r$ is the photon energy which crosses a surface of radius $r$. (We have $\varepsilon_{ph}(r) = dL_r/dM_r$.) The term $(acT^4)$ represents the photon energy density. The role of the opacity $\kappa$ as a moderator of the contraction is clearly expressed by the equation I–5. The virial theorem also gives us information on the mean values of the stellar parameters. The potential energy of the structure, at the moment when the radius is $R$, is given by the following (where $\bar{R}$ and $\bar{P}$ are defined by the integrals I-2 and I-6):

$$\Omega(R) = \int_0^R \frac{\rho G M_r \, dV}{r} = \frac{GM^2}{\bar{R}} = 3\bar{P}V \qquad \text{I-6}$$

whence $\bar{P} \propto M^2 R^{-4} \propto \bar{\rho}^{4/3} M^{2/3}$ and $d\bar{P}(t)\bar{P}/ = \frac{4}{3}d\bar{\rho}(t)/\bar{\rho}(t)$.

Now $dP/P = d\rho/\rho + dT/T$ for a perfect gas, hence we also have $d\bar{T}/\bar{T} = \frac{1}{3}d\bar{\rho}/\bar{\rho}$ and $\bar{T} \propto \bar{\rho}^{1/3} \propto 1/R$.

Thus, corresponding to each value of the parameter $R$ which characterizes the state of the contraction in hydrodynamic equilibrium, there is a value of $\Omega(R)$, $\bar{T}(R)$, $\bar{P}(R)$ and $\bar{\rho}(R)$.

The mean thermal energy per nucleon is given by $\frac{3}{2}k\bar{T}/\mu$ and is equal to the photon energy emitted since the beginning of the contraction.

Hence $-\Omega(R) = 3k\bar{T}/\mu$ gives the amount of energy per nucleon extracted from the gravitational "well".*

These results hold as long as the gas remains non-degenerate, which is the case as long as the number of particles ($n$) in a de Broglie thermal volume ($\lambda_T^3$) is less than unity:

$$n\lambda_T^3 \ll 1; \quad \lambda_T \equiv \frac{h}{(2\pi kTm)^{1/2}} \qquad \text{I-7}$$

By equation 1-7, $\lambda_T$ (electrons) $\simeq 40\lambda_T$ (protons), so that the electrons become degenerate long before the protons. In fact, stellar interiors where $\rho \simeq 10^5$ or $10^6$ gm/cm³ tend towards electronic degeneracy. Ionic degeneracy ($\leqslant 10^{10}$ gm/cm³) is probably never reached (except in the hypothetical neutron stars).

When condition I-7 no longer holds, the Pauli principle comes into play. This principle (applied to the fermions) restricts to two (spin) the population of each cell of volume ($h^3$) in the phase space. For complete degeneracy, all the boxes corresponding to energy lower than a maximal value ($E_F = p_F^2/2m$) are occupied while all the others are empty. We have thus for $n$ (the number of particles per unit volume):

$$n = \int_0^{p_F} \frac{2}{h^3}(4\pi p^2 \, dp) = \frac{8\pi}{3}\frac{p_F^3}{h^3} \quad \therefore \quad p_F \propto n^{1/3} \qquad \text{I-8}$$

We also have $\bar{p} = \int p \, dn(p) \Big/ \int dn(p) = \frac{3}{4}p_F$ so that $\bar{p} \propto n^{1/3}$. The energy density $u = (n \times$ average kinetic energy) and the pressure $P$ is $P \propto u$.

---

* We did not consider the effect of radiation pressure. This effect is negligible for nearly all stars in which we are interested.

In the non-relativistic case (N-R): $P \propto u \propto n\overline{p^2} \propto n^{5/3} \propto \rho^{5/3}$, and in the relativistic case (R): $P \propto u \propto n\overline{p} \propto n^{4/3} \propto \rho^{4/3}$.

These relations are strictly valid for a completely degenerate gas ($T = 0$). Otherwise we have rather, for instance in the N-R case:

$$P \propto \rho^{5/3}\left\{1 + a\left(\frac{kT}{E_F}\right)^2 + \ldots\right\} \qquad \text{I-9}$$

where $a$ is constant. Hence we see that as long as $kT \ll E_F$, the pressure is very weakly coupled to the temperature. This difference will be at the basis of the phenomenon called "stellar flash" (Lecture II).

### I–D: *Evolution in the P–ρ plane; limit values*

It is useful to consider here the behaviour of a gas of $n$ particles (fermions) in a log $P$–log $\rho$ diagram. Here, $P$ will be the total pressure including the radiation pressure: $P_{\text{rad}} \propto acT^4 \neq f(\rho)$. The full lines are the isothermals.

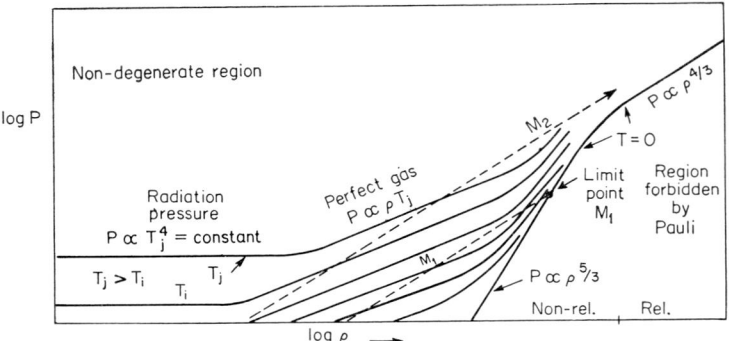

At low densities, the radiation pressure prevails. Then the gas pressure gets the upper hand ($P \propto \rho T$) so that the slope is unity. As we approach degeneracy the slope increases to $\frac{5}{3}$ (N-R case) or to $\frac{4}{3}$ (R case). The isothermals become asymptotically tangent to the isothermal $T = 0$ (complete degeneracy). The Pauli principle forbids access to the region on the right-hand side. In this diagram we trace the evolution of our star as it is given by the virial theorem: $\overline{P} \propto M^{2/3}\bar{\rho}^{4/3}$. We must distinguish here between two possible cases corresponding for instance to the paths of the masses $M_1$ and $M_2$ in the diagram.

Since the slope of the straight line traced by the point representing the average values $\bar{\rho}$ and $\overline{P}$ of the star $M_1$ is $\frac{4}{3}$, it is clear that because the isothermals change slope as we approach degeneracy: (1) there must be a maximal temperature (at that point the stellar evolution curve becomes tangent to the isothermal defined by this temperature); (2) any isothermal which is crossed once will be crossed another time, so that the temperature will go down again; (3) there exist a limit pressure and a limit density for this star (given by the point of intersection of the stellar evolution curve with the isothermal $T = 0$).

The thermal evolution of two stars $a$ and $b$ belonging to the first case ($M_{1b} > M_{1a}$) is shown in the diagram.

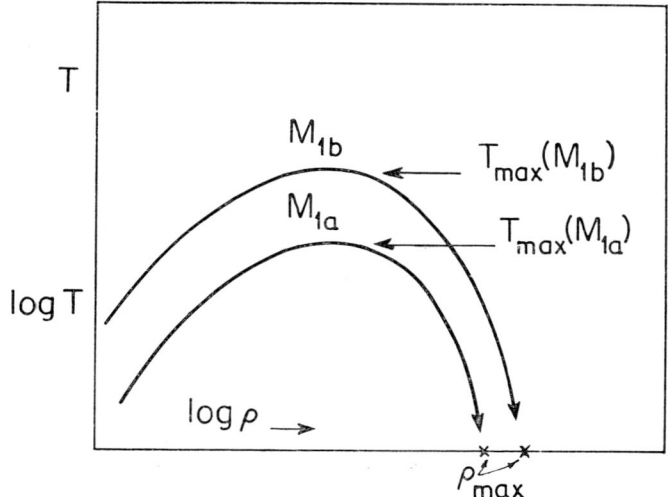

The maximum temperature of the heavier star will be higher than the maximum temperature of the lighter star; $(T_{\max}(M_{1b}) > T_{\max}(M_{1a})$. After cooling, the limiting densities will be such that
$$\rho_{\max}(M_{1b}) > \rho_{\max}(M_{1a}).$$
Then
$$\varepsilon_{ph} = \frac{dL_r}{dM_r} = \frac{P}{\rho^2}\frac{d\rho}{dt} - \frac{du}{dt} \Rightarrow 0$$
the star slowly dies out. At the limit:
$$\bar{P}_{gr} \propto \bar{\rho}^{4/3}M^{2/3} \propto \bar{\rho}^{5/3} \quad \therefore \; R \propto M^{-1/3}$$
Detailed computation gives ($M_\odot$ = mass of the sun):

| $M/M_\odot$ | $\log \rho_{\text{centre}}$ | $R/R_\odot$ |
|---|---|---|
| 0·2 | 5·4 | 1/50 |
| 0·8 | 7·2 | 1/100 |

These "petrified" objects exist in nature. They are the white dwarfs.

Thus the first case ($M_1$) contains the set of all stars whose final configuration ($T = 0$) contains no particles of relativistic energy ($p_F > mc$). The value of $p_F$ increases with the final density ($P_f \propto n^{1/3}$) and hence also with the mass ($P_f \propto M^{2/3}\bar{\rho}_f^{4/3} \propto \bar{\rho}_f^{5/3}, \therefore \bar{\rho}_f \propto M^2$). Consequently, to larger and larger masses there correspond more and more relativistic distributions. Now for such configurations, the isothermal $T = 0$ is parallel to the evolutive line. The contraction could thus, theoretically, continue indefinitely ($M_2$ in the diagram). However, other factors come into play. We will discuss this again later (Lecture III).

The limit mass between the two cases (called the Chandrasekhar limit mass) corresponds in practice to approximately $1{\cdot}5 M_\odot$.

### I–E: *Nuclear and weak interactions*

Continuing our study we now "switch on" together the nuclear and weak interactions. These two interactions, because of their short range, cannot come into play if the ions do not interpenetrate during a collision. At low temperatures, these interactions are actually held in check by the Coulomb repulsion. For two particles with masses $M_1$ and $M_2$ and charges $Z_1$ and $Z_2$, the probability of a collision in terms of the temperature is given by (Lecture II)

$$P(T) \propto \exp\left\{-\left(\frac{aZ_1^2 Z_2^2 \mu}{T}\right)^{1/3}\right\} \propto \left(\frac{T}{T_0}\right)^n \qquad \text{I–10}$$

where $a$ is a constant; $\mu$ is the reduced mass; $n = (aZ_1^2 Z_2^2 \mu/T_0)^{1/3}/3$ is an effective exponent (in the neighbourhood of a given temperature $T_0$). This exponent is in general very high (4 to 30 in certain cases): the probability increases very rapidly with temperature; when a certain temperature is reached, they are essentially "switched on"!

We distinguish three important effects of the nuclear reactions:

I–E–1) *Energy production*: for example the fusion of 4 hydrogens into helium releases 6·7 MeV per nucleon. The energy balance adds this new source to its credit:

$$\varepsilon_{ph} = \overbrace{\frac{P}{\rho^2}\frac{d\rho}{dt} - \frac{du}{dt}}^{\varepsilon_g} + \varepsilon_N \qquad \text{I–11}$$

Contraction continues until $\varepsilon_N \simeq \varepsilon_{ph}$ (see diagram). Then starts a period of nuclear fusion. The central variables $T_c$, $P_c$, $\rho_c$ remain more or less fixed. This period ends when the concentration of reacting particles

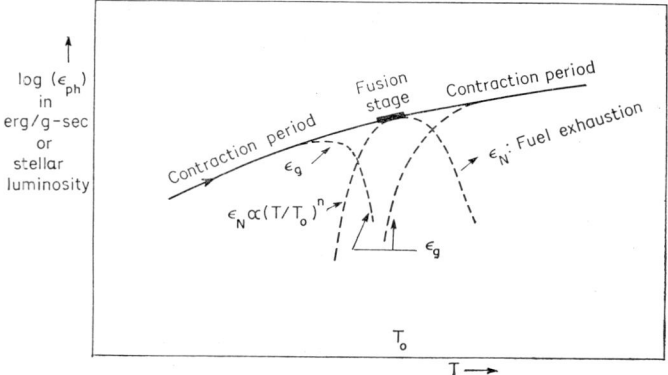

has strongly decreased. Thus during these periods the source of energy stops being gravitational and becomes nuclear. Instead of a macroscopic contraction of the star, accompanied, according to the virial theorem, by emission of energy extracted from the gravitational potential $\Omega_g$, we have multiple microscopic contractions (4 hydrogens into helium) accompanied by emission of gamma rays extracted from the nuclear potential.

The total energy emitted by the nuclear reactions during *the whole length* of a fusion stage is of the order of the MeV per nucleon, whereas the energy emitted by the gravitational contraction is of the order of the keV per nucleon. Since the *rate* of emission of this energy (the stellar luminosity) does not vary very much from one period to the next, we must expect that the length of the nuclear periods exceeds by far the length of the periods of gravitational contraction. It follows thus that the nuclear reactions considerably increase the life expectancy of a star. It is this fact which settled a famous dispute on the age of fossils (Cuvier), compared with the age of the sun (Kelvin).

I–E–2) *Certain nuclear reactions are accompanied by neutrino emission.* Unlike photons, neutrinos escape electromagnetic interaction. For neutrinos, the opacity $\kappa_\nu$ of the matter is essentially null. (Their mean free path in lead is of several light-years.) The neutrinos emitted at the centre of the star escape immediately. To be put to the debit of the star.

$$\varepsilon_g + \varepsilon_N = \varepsilon_{ph} + \varepsilon_\nu \qquad \text{I–12}$$

I–E–3) *Production of new elements (nucleosynthesis).* In general, the charges of these new elements are higher than those of the generating elements (e.g. $4H \to He^4$, $3He \to C^{12}$). At suitable temperatures these elements in turn become fuels. The more charged elements require higher temperatures before they can interpenetrate effectively (cf. the form of equation I–10). These temperatures are reached during the contraction periods which follow the exhaustion of the preceding fuel.

We now see the evolution pattern: gravitational and nuclear periods alternate. Evolution is accompanied, for the first, by an increase of the central density and, for the second, by an increase in the average mass of the atoms in the stellar nucleus (and also in the outer shell, as we will see later).

I–F: *The stages of thermonuclear fusion*

At present, one distinguishes between four periods of thermonuclear burning. Their characteristics as well as those of the gravitational periods which precede and follow them are described in the table. The calculations which lead to these results will be explained in Lecture II.

| $T_6$ ($10^6$ °K) | Total grav. energy emitted since the beginning | Reactions | Total nuclear energy emitted since the beginning | Limit mass | Photons | Neutrinos |
|---|---|---|---|---|---|---|
| | | | | | % | % |
| Grav. 0 → 10 | ~1 keV/n | | | | 100 | |
| Nucl. 10 to 30 | | $4H \to He^4$ | 6.7 MeV/n | 0.1$M$ | ~95 | ~5 |
| Grav. 30 → 100 | ~10 keV/n | | | | 100 | |
| Nucl. 100 to 300 | | $3He^4 \to C^{12}$, $4He^4 \to O^{16}$ | $\simeq 7.4$ MeV/n | 0.4$M$ | ~100 | |
| Grav. 300 → 800 | ~100 keV/n | | | | ~50 | ~50 |
| Nucl. 800 to 1100 | | $2C^{12} \to$ Mg, Ne, Na, Al | $\simeq 7.7$ MeV/n | 0.7$M$ | | ~100 |
| Grav. 1100 → 1400 | ~150 keV/n | | | | | ~100 |
| Nucl. 1400 to 2000 | | $2O^{16} \to$ S, Si, P | $\simeq 8.0$ MeV/n | ~0.9$M$ | | ~100 |
| Grav. 2000 → 5000 | ~400 keV/n | $\to$ Fe | $\simeq 8.4$ MeV/n | | | ~100 |

The first column represents the temperature ranges of each period. The masses considered here range from $0.1 M_\odot$ to $30 M_\odot$. The second column gives the total gravitational energy (per nucleon) emitted from the beginning of the contraction till the period under consideration. The third column identifies the nuclear reactions responsible for the corresponding period of fusion. The fourth column gives the total nuclear energy (per nucleon) emitted from the beginning of the "nuclear" contraction till the end of the period under consideration. Figs. I–2 and I–3 illustrate this table.

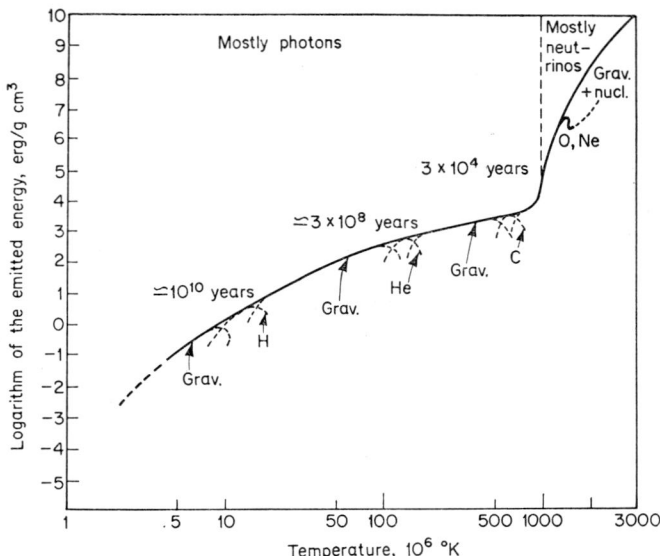

Fig. I–2. A very approximate evolution of the energy emitted by a star of about one solar mass whose internal temperature would reach some billions of degrees. The origin of the energy is identified by "Grav." during the periods of gravitational contraction and by the name of the nuclear fuel during the stages of nuclear fusion. The dotted lines represent the contributions of each of these mechanisms. In the left-hand side of the diagram, emission is mostly in the form of photons; in the right-hand side, in the form of neutrinos.

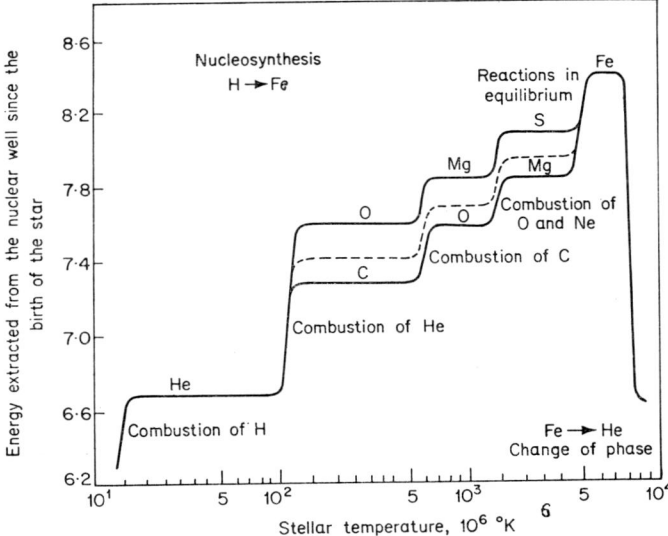

Fig. I–3. The nuclear energy, per nucleon, emitted since the beginning of the evolution, in terms of the central temperature. During the fusion stages, the curve rises rapidly (nearly isothermal processes).

Let us now consider again the effect of degeneracy on the stellar structure. For small masses, the existence of a limit temperature (Section I–D) shows that all stars will not go through the entire table; the evolution will reach the later stages only if this temperature is high enough. Some studies have shown recently that any mass smaller than $0 \cdot 1 M_\odot$ will never reach the stage of hydrogen burning. These stars will thus remain eternally composed of pure hydrogen. Helium burning will not start if $M < 0 \cdot 4 M_\odot$. In this case, the white dwarfs obtained upon cooling will have a nucleus of helium and an envelope of non-consumed hydrogen (the low surface temperature prevents the complete exhaustion of the hydrogen). If $M < 0 \cdot 7 M_\odot$, the carbon burning does not start. We then have a core of C (and O), a layer of He and an envelope of H. The limit mass for the oxygen burning is not yet known, it can be estimated as $\simeq 0 \cdot 9 M_\odot$. The diagram illustrates these results.

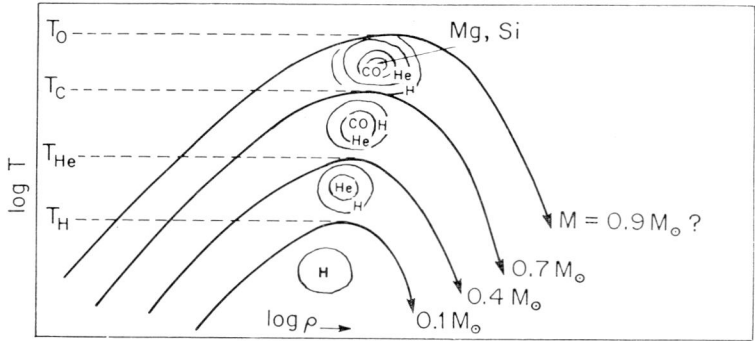

I–G: *Electromagnetic interaction coupled with weak interaction: pair formation; neutrinos*

In order to proceed further, we must go back a little for the time being and reconsider the effects of the electromagnetic interactions on the particles of our gas. With the increase in temperature, the density energy of the photons ($\propto acT^4$) has considerably increased. The average energy of these photons is $h\nu = 3kT$, and 4% of these photons have an $h\nu > 10kT$. Two new types of phenomena will appear in the star, the first (A) when a sufficient number of photons will go beyond the threshold $h\nu = m_e c^2 = 0 \cdot 511$ MeV, where $m_e$ is the mass of the electron; the second (B) when we will reach the region $h\nu = \Delta M_A c^2 \simeq$ several MeV, where $\Delta M_A c^2$ is the order of magnitude of the mass differences between the neighbouring nuclei. We will now consider these phenomena and briefly discuss their effect on the stellar gas.

The reaction $\gamma \to e^+ + e^- \to \gamma$ generates an equilibrium concentration of electrons and positrons. With a branching ratio of the order of $10^{-20}$ (determined by the ratio of the Fermi coupling constant and the electromagnetic coupling constant), pair annihilation gives $e^- + e^+ \to \nu + \bar{\nu}$ (instead of $e^+ + e^- \to 2\gamma$). However, these neutrinos immediately escape from the star, whereas the photons are absorbed and re-emitted a very large number of times before reaching the surface of the star.

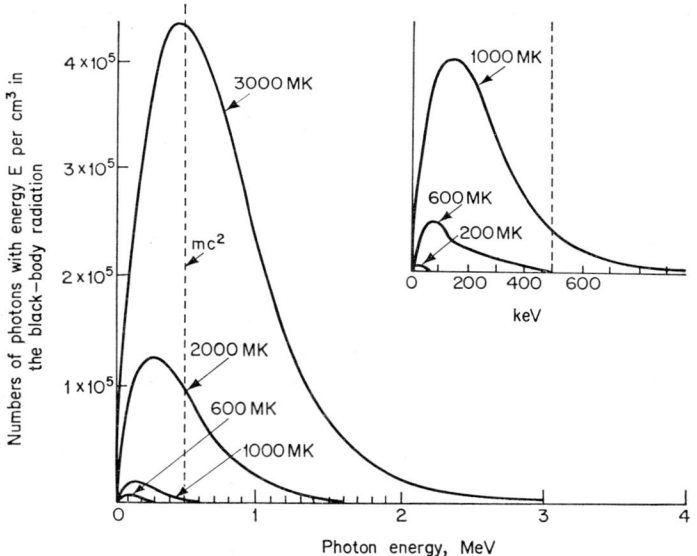

Fig. I-4. This diagram illustrates the (Planck) distribution of the photons with energy $E$ (MeV) in terms of the temperature (here MK = $10^6$ °K). We are interested here in the high-energy photons, in particular those which exceed $0.5 \text{ MeV} = m_e c^2 =$ rest mass of the electron.

In Lecture III, we will present, at least schematically, the theory of neutrino emission, and examples of calculation will be given. The logarithmic exponents $n = d \log \varepsilon_\nu / d \log T$ range from 9 to more than 20. In the neighbourhood of $T_6 = 500$ ($T_6 = T$ in $10^6$ °K), the (calculated) neutrino emission is greater than the photon emission (calculated or observed) of the brightest stars ($\varepsilon_\nu \gg \varepsilon_{ph} \simeq 10^5$ erg/gm/sec; for the sun $\varepsilon_{ph} = 2$ erg/gm/sec). In other words, at these temperatures we go from a photon emission mode to a neutrino emission mode. This new way of emitting energy allows a much more rapid dissipation of the stellar potential energy. The stellar stages will be characterized by $\varepsilon_N = \varepsilon_\nu \, (\gg \varepsilon_{ph})$, and the corresponding periods will be very much shortened. We should also mention that in degenerate media the neutrino emissions are much less intensive, and that the preceding remarks do not apply to the same extent.

After the oxygen-burning phase, the nuclear reserves are practically drained (there remain merely a few hundred keV per nucleon), and gravitational contraction becomes the only one in a position to meet the demands of the huge neutrino production ($\varepsilon_g = \varepsilon_\nu \simeq 10^{10}$ erg/gm/sec). By the virial theorem, the contraction increases the central temperature, which increases $\varepsilon_\nu$, which increases the contraction rate. It is obvious that the velocity of contraction cannot increase indefinitely without our reaching an unstable situation; when the velocity of contraction reaches the so-called free-fall velocity, hydrodynamic stability (represented by $dP/dr = -\rho G M_r / r^2$) is lost, and there is implosion.

I–H: *Electromagnetic interaction: nuclear photodisintegration*

The rates of photodisintegration will be studied in Section II–D. In the neighbourhood of $T_6 = 3000$ and over, quite a few elements disintegrate, emitting protons, neutrons and alphas, which are in turn captured by other nuclei, etc. A statistical equilibrium is progressively established between the abundances of the elements. This equilibrium favours the more stable elements (hence those which are at the bottom of the nuclear well) (Fig. I–5). This would be in particular Fe (especially 56), but also Mn, Co and Cr, Ni. On first analysis, the relative distribution of the isotopes is given by $n_i/n_j = \exp(-\Delta M_{ij}/kT)$, where $\Delta M_{ij}$ is the mass difference (per nucleon) between the isotopes $i$ and $j$, and $T$ is the temperature at the time when the phenomenon occurs.

At this time the structure of the star is in layers (like an onion). Around the iron nucleus, there is a layer of Si, P, S (products of the oxygen burning), a layer of C, O (helium burning), a layer of He (hydrogen burning) and finally an envelope of hydrogen.

If the temperature reaches $T_6 \simeq 5000$, the statistical equilibrium goes rapidly from Fe$^{56}$ to helium $\gamma + \text{Fe}^{56} \to 13\text{He}^4 + 4n$. The reaction is violently endothermic ($\Delta \Omega_N = -2\cdot 2$ MeV/nucleon), the nuclear energy is now written to the debit of the star. Several studies have shown that the hydrostatic equilibrium does not resist this change.

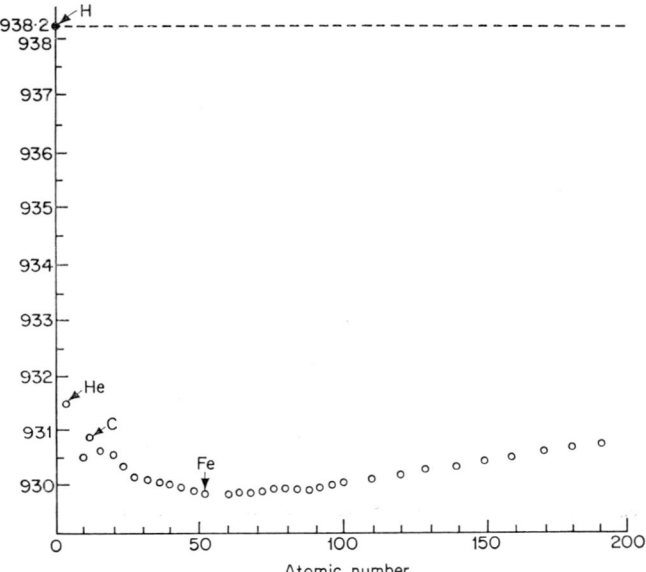

Fig. I–5. The mass per nucleon ($Mc^2/A$) of the stable nuclei (the famous nuclei-stability curve). Note the minimum in the neighbourhood of iron.

Whether it be by neutrino emission or by photodisintegration (it is not yet known which will occur first), the action of the E.M. interactions (accompanied by W.I. and the N.I.) inevitably leads our stellar structure to catastrophe.

The implosion would bring to the centre of the star the stellar layers composed of potential fuels (H, He, C, O, etc.). The rapid burning of these fuels would cause an explosion which would throw out again part of the mass accumulated in this way. This explosion is often associated with the appearance in the sky of the "supernovae" (Lecture III). The rejected matter contains the nucleosynthetic product of the stellar reactions. The interstellar gas is enriched with elements heavier than hydrogen. A fraction of the stellar mass probably falls back onto itself, and eventually becomes a dwarf comparable to the one we have met earlier.

It seems it is now admitted that a star can also lose part of its mass during its stable evolution by a more or less continuous surface ejection (stellar wind). The planetary nebulae are an example of this.

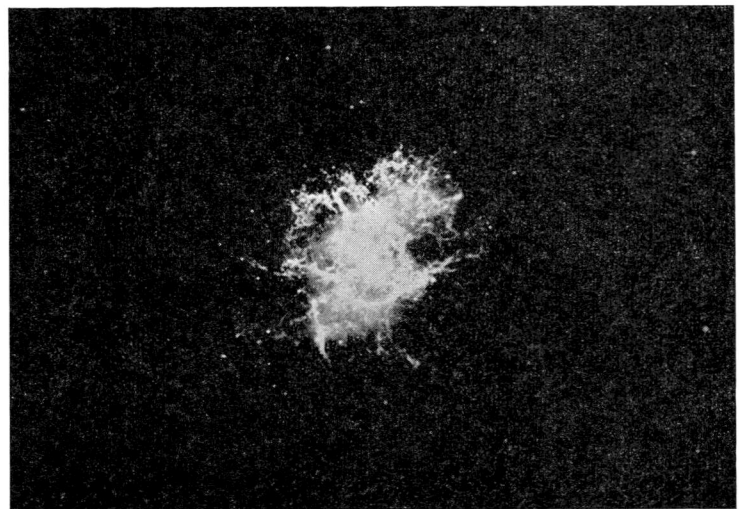

Fig. I-6. The famous Crab nebula. It is the remnant of a star (supernova) which exploded nine hundred years ago. According to the Chinese astronomers who observed it, this event promised abundant crops in a very near future. The interpretation holds true, except for the time factor . . .

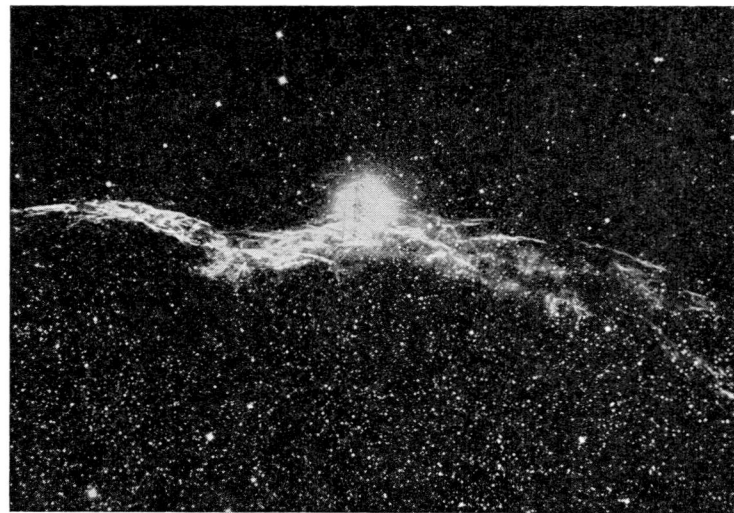

Fig. I-7. The remnant of a much older explosion (several tens of thousands of years). The enriched gas is finishing to mix with the interstellar gas.

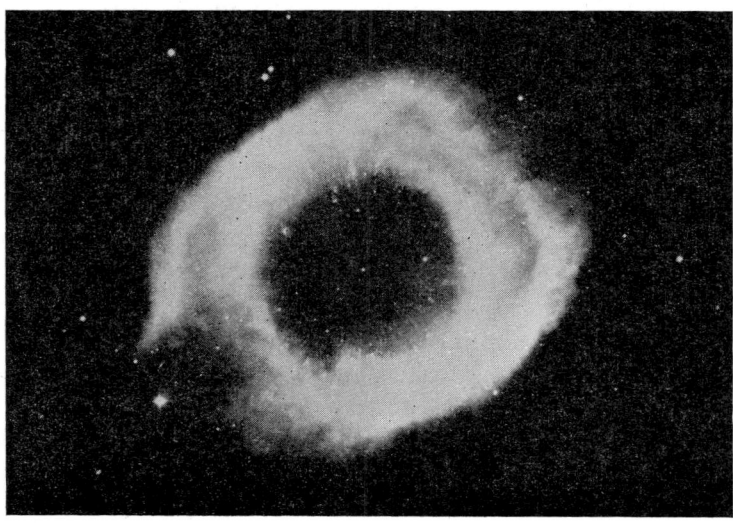

Fig. I–8. A planetary nebula. The luminous crown consists of the gas ejected from the central star. The phenomenon continues non-violently for a long period of time.

### I–I: *Final energy balance*

The G.I. and N.I. have generated the energy necessary for stellar evolution. If we consider the evolution from the beginning till the period immediately preceding the implosion, the $\Delta\Omega_g$ is of about 0·5 MeV/nucleon and the $\Delta\Omega_N$ is of about 8·4 MeV/nucleon, so that a total of $\simeq 9$ MeV/n have been emitted. The E.M.I. and W.I. have dissipated this energy. The E.M.I. (especially in the beginning, $T_6 \lesssim 500$) dissipated about 7 MeV/nucleon, while the W.I. dissipated $\simeq 2$ MeV ($T_6 \gtrsim 500$).

### I–J: *Galactic evolution; nucleosynthesis; stellar populations*

The formation of a certain number of dwarfs (mass of matter closed upon itself and unaffected by the surrounding activity) resulted from the activity of the G.I. The fraction of galactic mass which is closed up in this way must increase with time. Young star clusters should contain less dwarfs than the older clusters. In practice, observation of dwarfs is difficult, but other types of observations suggest that this is indeed the case.

Nucleosynthesis followed from the activity of the N.I. The experimental data is illustrated in Fig. I–9. In the figure, we see the distribution of elements in the solar system (earth, meteorites, sun). The elements can be divided into four important groups: the hydrogen peak ($A < 40$ (except $A = 6, 7, 9, 10, 11$)), the iron peak ($40 < A < 65$), the heavy elements ($A > 65$), and the group Li, Be, B.

We have assigned the first group to the nucleosynthetic result of the nuclear reactions responsible for the stellar energy. The corresponding rates of reaction are governed by the Coulomb factors, hence by the charge. The strongly decreasing slope of $n(A)$ in terms of $A$ (hence of $Z$) illustrates hence the inhibiting effect of the electrostatic repulsion; the high temperatures necessary for fusion of the more charged elements are more difficult to obtain, hence more rare.

We have assigned the second group to the nuclear processes in thermodynamic equilibrium (photodisintegration and recombination) $T_6 \simeq 3000$ or $4000$. Here the charge no longer plays any role; it is the

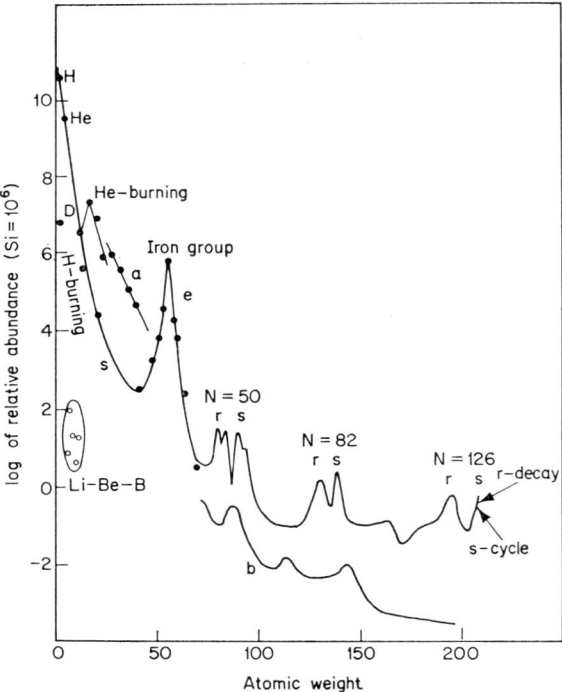

Fig. I–9. The so-called "cosmic" abundance curve of the elements (from Suess-Urey).

stability of the nuclei that matters. The most stable one (Fe) occurs most and the slope of the peak ($\Delta M_{ij}/kT$) reflects the value of the ambient temperature.

The last two groups (much more rare) are not directly implicated in the energy evolution of a star. We will assign (Lecture IV) the synthesis of the heavy elements to the capture of neutrons by elements of the iron peak. These neutrons are generated by certain secondary reactions which occur during the stellar evolution, but do not influence it in any way. As for the fourth group (Li, Be, B), we now have good reason to believe that it is the result of nuclear reactions on the surface of the stars.

The curve I-9 represents in fact the distribution of the isotopic abundances in the gas which formed the sun (the nucleosynthetic carrier of the sun is not very advanced; it has converted in its centre roughly 5% of its initial hydrogen into helium). The age of the sun is roughly $5 \times 10^9$ years, so that the curve reflects the state of the galactic gas at this time.

The methods of evaluation of the age of stars will be discussed in Lectures II and V. According to their age, stars are grouped in Populations I and II, Pop. I being the youngest. We then subdivide these two groups into extreme, young, intermediary and old Pop. I, then intermediary and extreme Pop. II. The ages are not well defined but range from $10^6$ to $10^{10}$ years.

For many of these stars we can draw a more or less rudimentary figure (I-9). If we group together the elements $Z > 2$, we clearly see a correlation between the age of the star and the abundance of these elements. For very old stars (extreme Pop. II), these elements represent less than 0·003 of their superficial composition; for the older Pop. I, more or less 0·02, and for the young Pop. I, $\simeq 0\cdot04$. This result remains one of the most solid supports of stellar nucleosynthetic theory.

BIBLIOGRAPHY

The bibliography contains, on the one hand, a series of books which present a synthesis of the subject and, on the other hand, a series of works sufficiently recent not to appear, in general, in review articles. The bibliography does not claim to be complete, nor to give everyone the credit he deserves. We only wish to give some food for thought wanting to go beyond these notes. A very complete bibliography has recently appeared in Poland. It can be obtained:

B. KUCHOWICZ, "Nuclear Astrophysics; A Bibliographical Survey", Review Report no. 8, 9, 10. Nuclear Energy Information Centre.

Write to: Polish Government Commissioner for Use of Nuclear Energy
Palace of Culture and Science
*WARSAW*
Poland

Also:

E. LANGER, M. HERZ, J. P. COX, "Recent Work on Stellar Interiors: A Bibliography of Material Published Between 1958 and Mid-1966", J.I.L.A. Report No. 88, University of Colorado, Boulder, Colorado.

*Bibliography*

The bibliography for this lecture contains several rather general works which also cover several of the later lectures. The titles will *not* be repeated in the other bibliographies.

To begin with, two classics:

A. S. EDDINGTON, "The Internal Constitutions of the Stars", Cambridge University Press (1930)
S. CHANDRASEKHAR, "Stellar Structure", University of Chicago Press (1938)

Then works containing a general synthesis of Astrophysics:

E. SCHATZMAN, J. C. PECKER, "Astrophysique Générale", Masson (1959)
F. KAMENETSKII, "Physical Processes in Stellar Interiors", Insrael Programme of Scientific Translation
V. A. AMBARTSUMYAN, "Theoretical Astrophysics", Pergamon Press, London (1959)

Interiors and stellar evolution are, in particular, dealt with in:

M. SCHWARZSCHILD, "Structure and Evolutions of the Stars", Princeton University Press (1958) (a splendid book)
D. H. MENZEL, P. L. BHATNAGAR, H. K. SEN, "Stellar Interiors", Chapman & Hall, London (1963)

Nuclear Astrophysics and nucleosynthesis are treated in:

E. M. BURBIDGE, G. R. BURBIDGE, W. A. FOWLER, F. HOYLE, "Synthesis of Elements in Stars", *Rev. Mod. Phys.* **29**, 547, 1957 (for a long time, the bible of nucleosynthesis)
A. G. W. CAMERON, "Nuclear Astrophysics", Goddard Institute for Space Studies (1964)
G. BURBIDGE, *Ann. Rev. of Nucl. Sciences*, **12**, 507, 1962

Some series devoted to these problems:

"Advances in Astronomy and Astrophysics", (Kopal), Academic Press (2 volumes)
"Stars and Stellar Systems" (McLaughlin, Aller and Middlehurst), University of Chicago Press (eventually in 10 volumes)
"Handbuch der Physik", vols. L–LIV, Springer Verlag, Berlin
Congrès et Colloques de l'Université de Liège, Université de Liège
Transactions of the International Astronomical Union
Symposiums of the Goddard Institute for Space Studies, New York City, N.Y.

Some reports of Summer courses:

Varenna (1962), "Star Evolution" (L. Gratton), Academic Press, New York City
"Lectures in Theoretical Physics", Vol. VI, University of Colorado Press (1963)

# II

## The role of charged particles

II–A: *The mapping between the Z–N plane and the H–R plane*

WE have seen how the life of a star is composed of alternating stages of gravitational evolution and nuclear evolution; because of the nuclear reactions, the contraction is, on several occasions, interrupted. The length of these interruptions depends in part on the nature of the nuclear fuel in operation.

During these nuclear stages, the evolution of the macroscopic parameters of a star (luminosity $L$, surface temperature $T_e$, radius $R$) is slowed down very much. This phenomenon can be observed by means of statistical studies of the stars, for example by means of the Hertzsprung–Russell (H–R) diagram, which we will consider again later. On the other hand these stages are also characterized by a change in the

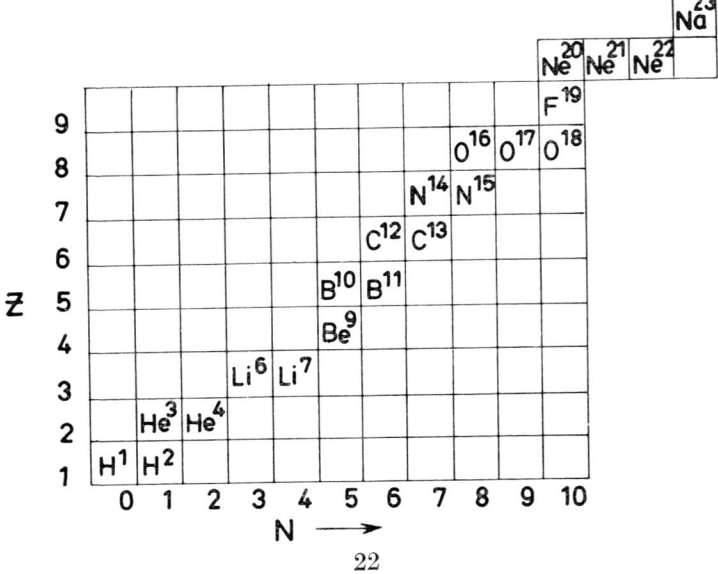

chemical composition of the star (hence microscopic). In order to analyse the evolution of the chemical composition, it is useful to consider the Z–N plane which groups all the elements (and isotopes) in terms of their number of protons (Z) and of their number of neutrons (N). In this lesson we will study the correlation between the chemical composition of a star and its macroscopic parameters during evolution. In mathematical terms, we are going to develop a mapping theory between the Z–N plane and the H–R plane. This mapping depends first of all on the rate of nuclear interaction between the chemical components.

## II–B: *Nuclear reactions: generalities*

Let there be in 1 cm³, $n_1$ particles of type 1 (mass $A_1$, charge $Z_1$) with a Maxwell–Boltzmann (M.B.) velocity distribution $N(\vec{v_1})$ (as long as $\rho < 10^{10}$ gm/cm³, the gas of nuclei is not degenerate); let there be also $n_2$ particles of type 2 (different from 1) under the same conditions. It can readily be shown that the distribution of relative velocities $v = |\vec{v_1} - \vec{v_2}|$ is also M.B. (provided we use the reduced mass $\mu = A_1 A_2/(A_1 + A_2)$ in the formula).

$$N(v) = \left(\frac{2}{\pi}\right)^{1/2} \left(\frac{N_1 N_2 v^2 \mu^{3/2}}{(kT)^{3/2}}\right) \exp\left\{-\left(\frac{\mu v^2}{2kT}\right)\right\} \qquad \text{II–1}$$

$N(v)$ is here the number of pairs of particles with relative velocity $v$; the total number of pairs is $N_1 \cdot N_2$ (per cm³).

The number of nuclear reactions between 1 and 2 (per cm³ and per sec) is given by

$$I_{1,2} = \int_{v_s^-}^{\infty} N(v) P(v) \, dv \qquad \text{II–2}$$

where $P(v)$ is the reaction probability between two particles of relative velocity $v$; $v_s$ is the velocity corresponding to the threshold energy $E_s$ if the reaction is endothermic. (For an exothermic reaction, $v_s = 0$.)

The probability $P(v)$ is related to the cross-section $\sigma(v)$ by

$$P(v) = v\sigma(v) \qquad \text{II–3}$$

*Useful concepts:*

$$\langle \sigma v \rangle_{1,2} = \frac{\int_{v_s}^{\infty} N(v) P(v) \, dv}{\int_0^{\infty} N(v) \, dv} \qquad \text{Probability of reaction per unit pair} \qquad \text{II–4}$$

$P_{2,1} = N_1 \langle \sigma v \rangle_{1,2}$    Mean life of a particle 2 amongst $N_1$ particles 1. (Note the order of the indices).

$t_{2,1} = \dfrac{1}{P_{2,1}}$

$N_1 = \dfrac{\mathcal{N} \rho X_{A_1}}{A_1}$    (where $\mathcal{N} = 6 \cdot 02 \times 10^{23}$ and $\rho$ is the density). This equation defines $X_{A_1}$, the fractional density per mass of the element 1.

$\varepsilon_{1,2} = \dfrac{I_{1,2} Q_{1,2}}{\rho}$    Rate of production of energy per *gram-second* if $Q_{1,2}$ is the amount of energy emitted per reaction.

$n = \dfrac{T}{\varepsilon} \left( \dfrac{d\varepsilon}{dT} \right)_{T=T_0}$    And finally, the logarithmic derivatives which are the effective exponents of the energy with respect to the temperature and the density in the neighbourhood of $T_0$ and $\rho_0$

$m = \dfrac{\rho}{\varepsilon} \left( \dfrac{d\varepsilon}{d\rho} \right)_{T=T_0}$

$\varepsilon = \varepsilon(T_0, \rho_0) \left( \dfrac{T}{T_0} \right)^n \left( \dfrac{\rho}{\rho_0} \right)^m$

The factor $n$ can be very large (of the order of 30 to 40); this explains the sudden switching on of the nuclear reactions (cf. Figs. I–2 and I–3).

The case when the gas reacts with itself ($2 = 1$) is readily obtained; we replace $N_1 \cdot N_2$ by $N^2/2$ (the number of pairs of identical particles) in the expression for $N(v)$ (eqn II–1).

## II–C: *Nuclear reactions: charged particles*

Stellar atoms are in general ionized and bathe in a "sea" of free electrons. These electrons are not uniformly spread out but surround the nuclei with a sort of spherical cloud which affects the potential (eqn II–5). In first approximation, we neglect this effect: we will come back to it later (II–F).

Consider two particles with charge $Z_1$ and $Z_2$, separated by a distance $r$. The electrostatic potential energy is given by

$$U_r = \dfrac{Z_1 Z_2 e^2}{r} \qquad \text{II–5}$$

The probability of a nuclear reaction can be written as the product of the probability ($P_{\text{coul}}$) that the two particles approach one another at a distance $r = R$ (where $R$ is the minimum radius of interaction) and the probability ($P_{\text{nucl}}$) that the particles in contact react in a nuclear fashion.

The first stage is illustrated in the diagram (the penetration of the Coulomb barrier). We wish to calculate the probability that the particles with relative energy $E$ approach by a "tunnel" effect (since in Astrophysics $E \ll B$ almost always) up to $R$. The wave function of the

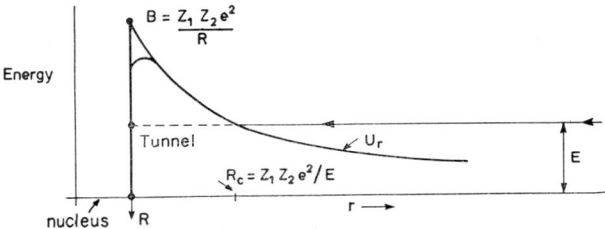

system is a superposition of the regular solutions $F_l(r)$ and the irregular solutions $G_l(r)$ of the Schrödinger equation with Coulomb potential. Then the tunnelling probability is given by ($E \ll B$):

$$P(r = R) \propto \{F_0^2(R) + G_0(R)^2\}^{-1} \propto e^{-2\pi\eta}E^{-1/2}$$

(where $\eta = Z_1 Z_2 e^2 / \hbar v$).

We still know so little about nuclear physics that we cannot, in general, calculate the term $P_{\text{nucl}}$. It must be obtained experimentally.

We can distinguish three possible cases:

II-C-1) *Non-resonant rate*. If the factor $P_{\text{nucl}}$ is more or less independent of the energy of the incident particles (the exact conditions will be specified later), then:

$$\sigma = \frac{P_{\text{coul}} P_{\text{nucl}}}{v} = \frac{S e^{-2\pi\eta}}{E} \qquad \text{II-6}$$

The factor $S$, called the "astrophysical factor", measures the probability of the purely nuclear reaction. It is experimentally determined by means of the inverse formula $S = \sigma E e^{2\pi\eta}$ ($\sigma$ is usually measured in barns (1 b = $10^{-24}$ cm$^2$) and $E$ in MeV, so that $S$ is in MeV-barns), measuring $\sigma_{\text{exp}}$ for several $E$.

In this case (we define $2\pi\eta = a/E^{1/2}$):

$$\langle \sigma v \rangle = \int n(E)\, v\sigma(E)\, dE \propto S \int_0^\infty \exp\left(-\frac{E}{kT} - \frac{a}{E^{1/2}}\right) dE \qquad \text{II-7}$$

In the diagram, the terms $n(E)$, $v\sigma(E)$ and their products are represented in terms of the energy:

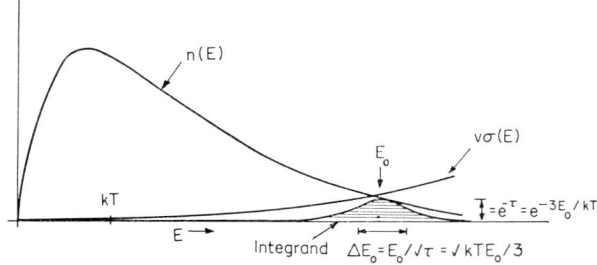

The integrant has a peak whose maximum occurs for an energy $E_0$ (called the Gamow energy) and its width is $\Delta E_0$. This peak contains the most likely candidates for nuclear reactions.

By the second term of equation II–7, the maximum of the peak corresponds to $a/2E^{3/2} = 1/kT$, hence $E_0 = (\frac{1}{2}akT)^{2/3}$. The height of the peak at $E_0$ is $\exp(-3E_0/kT) = e^{-\tau}$, and its full width at mid-height is $\Delta E_0 = E_0/\sqrt{\tau}$.

An approximate value of the integral is given by (height × width) $= \Delta E_0 e^{-\tau}$. The term $\langle \sigma v \rangle$ depends mostly on $S$ (nuclear factor) and on $e^{-\tau}$ (Coulomb factor). In the table, the energy $E_0$, the term $\log S$ and the term $\log(e^{-\tau})$ are given for a certain number of nuclear reactions in terms of the temperature ($T_i$ represents the temperature in $10^i$ °K).

| $Z_1 \rightarrow$ = | 1 | | | 2 | | | 6 | | |
|---|---|---|---|---|---|---|---|---|---|
| $Z_2 \downarrow$ | $T_6=15: kT=1\cdot 3$ keV | | | $T_6=100: kT=86$ keV | | | $T_6=1000: kT=860$ keV | | |
| | $E_0$ keV | $\log S$ keV-b. | $\log(e^{-\tau})$ | $E_0$ keV | $\log S$ keV-b. | $\log(e^{-\tau})$ | $E_0$ keV | $\log S$ keV-b. | $\log(e^{-\tau})$ |
| = 1 | 5·9 | −21·5 | −6·10 | | | | | | |
| 6 | 23·9 | 0·1 | −24·0 | 215 | 2·5 | −32 | 2200 | 20·3 | −33 |
| 16 | 47 | ≃ 3·0 | −47·5 | 415 | 16 | −63 | 4200 | ≃ 35 | −62 |
| 26 | 65·0 | | −66 | 500 | 22 | −75 | | | |

The temperature unit ($T_6$) is a million of degrees K.

We see from the table that for a given fuel (1), the probability of a reaction with a fuel (2) decreases very rapidly with the charge of the second fuel, and, except for the case ($Z_i = 1$, $Z_2 = 1$ and $Z_1 = 1$, $Z_2 = 6$), the increase of $S$ does not compensate for the decrease of $e^{-\tau}$.

This phenomenon explains the presence of well-defined nuclear stages instead of an uninterrupted sequence of successive fusions.

A better approximation for the integral is given by the substitution

$$\int \exp\left(-\frac{E}{kT} - \frac{a}{E^{1/2}}\right) dE \simeq e^{-\tau} \int \exp\left\{-\left(\frac{E - E_0}{\frac{1}{2}\Delta E_0}\right)^2\right\} dE \quad \text{II-8}$$

which allows us to evaluate numerically

$$\frac{P_{2,1}}{\rho X_1} = A \exp\left(-\frac{B}{T^{1/3}}\right) T^{-2/3} \quad \text{II-9}$$

In the tables of thermonuclear reaction rates, $A$ and $B$ of equation II-9 are generally given for astrophysical purposes.

The determination of $A$ obviously depends on the experimental measure of $S$. If $S$ exhibits experimentally a *weak* energy dependence, we replace it by $[S_0 + (dS/dE)_0 E_0]$ where $[dS/dE]_0$ is experimentally determined.

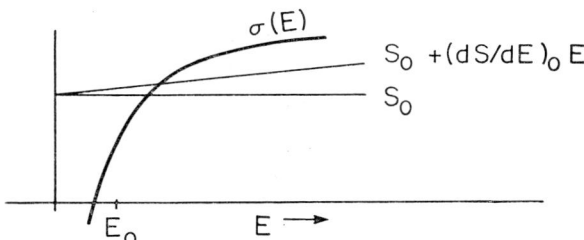

The behaviour of real nuclei is illustrated by the energy diagram of a typical case $O^{16}$ (Fig. II–1); they have excited levels $E^*$. These levels have a width $\Gamma$ related to their length $t$ by $t = \hbar/\Gamma$. This width can be subdivided into a sum of widths corresponding to the probability of de-excitation by emission of an $\alpha$, $\gamma$, $n$, etc. ($\Gamma = \Gamma_\alpha + \Gamma_\gamma + \Gamma_n + \ldots$). The nuclear probabilities ($P_{\text{nucl}}$) of reaction (for instance $C^{12} + He^4 \to O^{16} + \gamma$) do not depend very much on the energy, except in the neighbourhood of the excited levels (resonances are observed). We define the resonance energy $E_r = E^* - Q$ (see the diagram at the foot of page 28).

II–C–2) *Resonant rate.* In the neighbourhood of a resonance, the cross-section experimentally assumes the Breit-Wigner (B.W.) form:

$$\sigma_{\text{B.W.}} = \frac{\pi \lambda^2 w \Gamma_e \Gamma_s}{(\frac{1}{2}\Gamma)^2 + (E - E_r)^2} \quad \text{II-10}$$

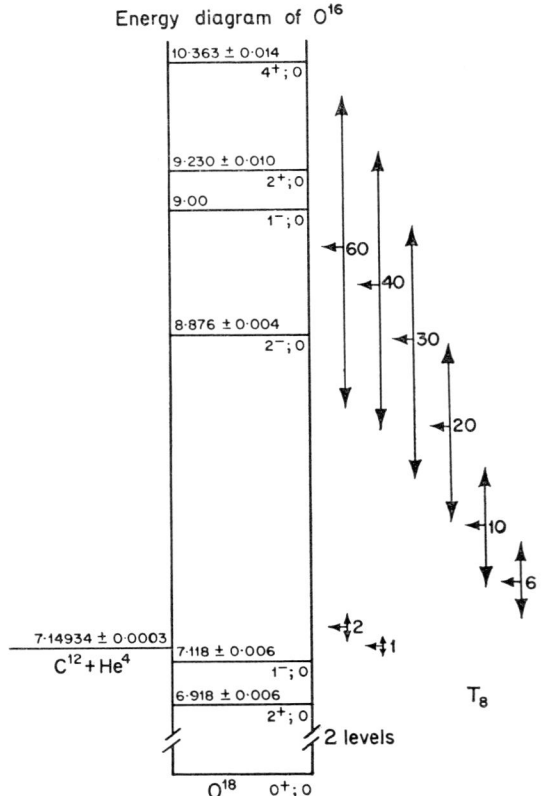

Fig. II-1. The energy diagram of $O^{16}$ and the threshold of the reaction $C^{12} + He^4$. The position (horizontal arrow) and the width (vertical arrow) of the Gamow peak of this reaction are schematized on the right-hand side of the diagram for various temperatures in units of $T_8 = 10^8 \,°K$.

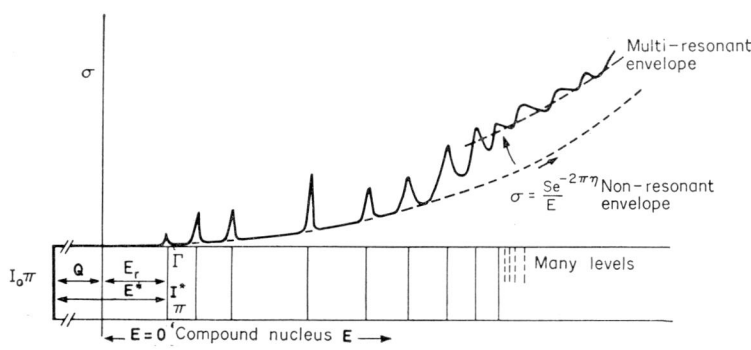

$\Gamma_e$ is the width for the channel of formation of the excited state; $\Gamma_s$ is the exit channel; $w$ is a statistical factor;

$$w = \frac{2I^* + 1}{(2I_0 + 1)(2S_0 + 1)} \quad \text{and} \quad \lambdabar = \frac{\hbar}{mv}$$

We assume $\Gamma_e = \Gamma_c$ (charged particle) and $\Gamma_s = \Gamma_\gamma$, then:

$$\Gamma_c = 2\gamma^*(kR)P_l$$
$$P_l = \{G_l^2(R) + F_l^2(R)\}^{-1} \propto \frac{e^{-2\pi\eta}}{E^{1/2}} \qquad \text{II-11}$$

where $k = 1/\lambdabar$, $R = $ radius of nuclear interaction, $\gamma^*$ is called the reduced width. Like $S$, it represents a purely nuclear factor.

In almost all cases we have $\Gamma = \Gamma_e + \Gamma_s + \ldots \ll \Delta E_0$. Then we only consider the contribution due to resonance and we get:

$$\langle \sigma v \rangle_r \simeq \frac{n(E_r)v(E_r)}{\int n(E)\,dE} \int \sigma(E)\,dE = w\lambda_T^3 \exp\left(-\frac{E_r}{kT}\right)\frac{\Gamma_e \Gamma_s}{\Gamma \hbar} \qquad \text{II-12}$$

where $\lambda_T = \dfrac{h}{(2\pi\mu kT)^{1/2}}$

This is the so-called resonant approximation (or better, the contribution due to the resonance).

Experimentally, one can often obtain directly

$$\int \sigma(E)\,dE = \frac{w\Gamma_e \Gamma_s}{\Gamma} 2\pi^2 \lambdabar^2 \qquad \text{II-13}$$

by means of a beam directed on a thick target.

The reaction rate can then be written as

$$\frac{P_{2,1}^{\text{res}}}{\rho X_1} = F\,e^{-G/T} \qquad \text{II-14}$$

where $F$ contains the widths $\Gamma$, and $G$ the energy $E_r$.

Far from the resonances, $\sigma_{\text{B.W.}}$ can be written as

$$\sigma_{\text{B.W.}} = \left(\frac{\pi \lambdabar^2 E}{E}\right)\left\{\frac{w\Gamma_e\,e^{2\pi\eta}}{(E-E_r)^2}\right\}\Gamma_s\,e^{-2\pi\eta} = \left(\frac{S}{E}\right)e^{-2\pi\eta} \qquad \text{II-15}$$

where $S$ is more or less constant if $\Gamma \ll |(E-E_r)|$. Hence we have again the previous case (non-resonant) with $S$ determined by the equation II-15.

In the case where the Gamow peak contains no resonance, the non-resonant rate is generally the largest. If the peak contains one or more resonances, the situation is reversed. In both cases, the approximation II–9 and the approximation II–14 hold, respectively.

II–C–3) *Multiresonant reactions*. These happen when there are several resonances in the Gamow peak.

In this case, we can add up the rates II–14 if we know all the parameters $\Gamma$, $E_r$, etc. We can also use a more generally applicable statistical method. This method introduces the concept of "strength function".

We first define in an interval ($\Delta E \gg \Gamma$) an average value $\bar{\sigma}$ which can be measured experimentally by means of an instrument with poor resolution ($\Delta E$):

$$\bar{\sigma} = \int \frac{\sigma\, dE}{\Delta E} = \sum_{\text{res}} \frac{2\pi^2 \lambda^2 w \Gamma_e \Gamma_s}{\Gamma \Delta E} \qquad \text{II–16}$$

If $\Gamma_e < \Gamma_s$ and $\Gamma = \Gamma_e + \Gamma_s$ (which will generally be the case for energies lower than the Coulomb barrier), then $\Gamma_e \Gamma_s / \Gamma \simeq \Gamma_e$ and $\bar{\sigma} = (S/E)\, e^{-2\pi\eta}$, where $S = (2\pi^2 \lambda^2 E) \Sigma\, (\Gamma_e\, e^{2\pi\eta} w / \Delta E)$ is proportional to the "strength function".

Multiresonant rate is thus of the same form as non-resonant rate. This is natural since at that time the average behaviour of the nucleus is relatively independent of the energy.

The average behaviour can also be evaluated by means of the optical model of the nucleus. This model describes the phenomena of scattering and absorption (capture reactions) by means of a complex nuclear potential of the form

$$V(r) = \frac{V_0 + iW_0}{1 + \exp\{(r - R_0)/a\}} \qquad \text{II–17}$$

The behaviour of a large number of nuclei is well represented by a suitable choice of the parameters $V_0$, $W_0$, $R$, $a$. In general, these parameters vary in a rather continuous fashion with $A$ and $Z$ of the target, which permits a certain form of interpolation for unmeasured reactions.

For the multiresonant capture rate, this model gives a good representation of the form $\sigma = S/E\, e^{-2\pi\eta}$ and allows us to evaluate $S$ directly (cf. Figs. II–2 and II–3).

The rate of multiresonant reaction will thus be expressed by

$$\frac{P}{\rho X_1} = A \exp\left(-\frac{B}{T^{1/3}}\right) T^{-2/3}$$

where $A$ now reflects the intensity of the strength function for the corresponding reaction.

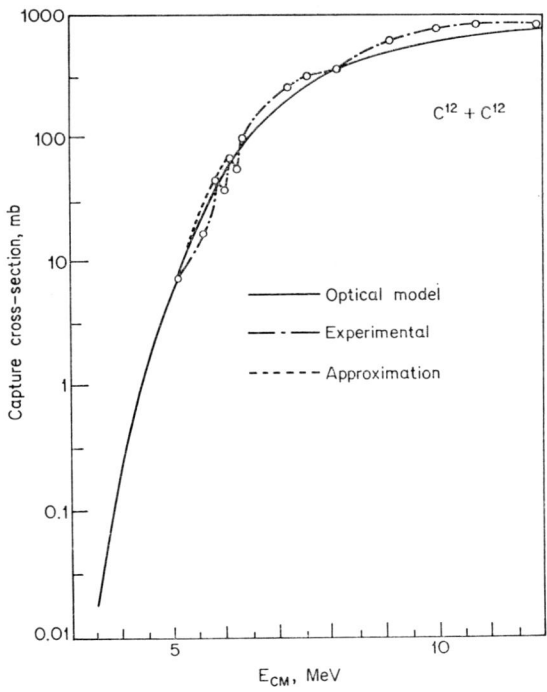

Fig. II–2. The excitation curve of the reaction $C^{12} + C^{12}$ as a function of the energy. Besides the experimental curve we see the result of a calculation with an optical model of the nucleus, and also the curve of equation II–15 with suitable parameters (it coincides with the full line until $E \sim 5$ MeV). We can see clearly the "non-resonant" behaviour of the multiresonant reactions.

II–D: *Photodisintegration rates*

At rather high temperatures ($T_6 \gtrsim 500$) we must take the photodisintegration reaction (destruction of a nucleus by absorption of gamma rays) into account. We write:

$$C + \gamma \to C^* \to A + B$$

where $C^*$ represents the nucleus $C$ in an excited state. The disintegration rate is calculated in three steps.

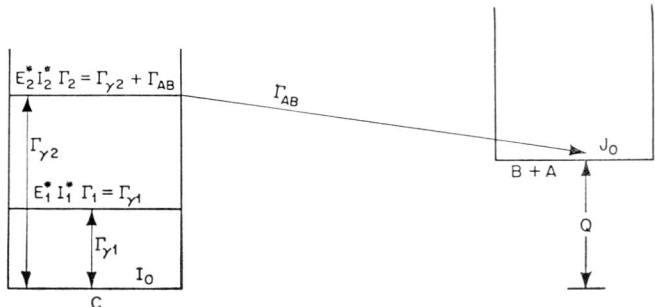

(A) If $E_1^* < Q(\Gamma_{1AB} = 0)$, the relative population of an excited state with respect to the ground state is given by

$$p_1 = \exp\left(-\frac{E_1^*}{kT}\right)\left(\frac{2I_1^* + 1}{2I_0 + 1}\right) \qquad \text{II-18}$$

(B) If $E_2^* > Q$, we must correct this term; a fraction $\Gamma_{2AB}/\Gamma_2$ of the

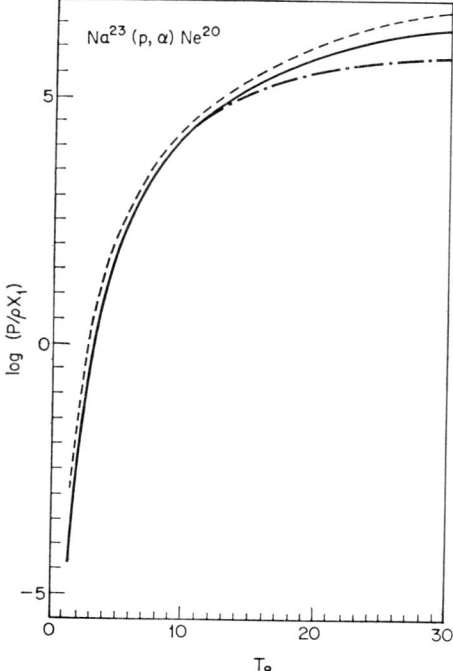

Fig. II–3. The rate of thermonuclear reaction of $Na^{23}(p, \alpha)Ne^{20}$ as a function of the temperature. The full line represents the sum of the resonant rates, whereas the dotted line (----) represents the multiresonant rate obtained with the optical model of the nucleus.

nuclei in the excited state are destroyed by the reaction $C \to A + B$

$$p_2 = \exp\left(-\frac{E_2^*}{kT}\right)\left(\frac{2I_2^* + 1}{2I_0 + 1}\right)\left(\frac{\Gamma_{\gamma_2}}{\Gamma_2}\right) \qquad \text{II-19}$$

(C) The probability of disintegration is now simply the product of the probability of disintegration of an excited nucleus, $\Gamma_{2AB}/\hbar$, times the probability of finding the nucleus in this excited state $p_2$:

$$P_{\text{photo}} = \exp\left(-\frac{E^*}{kT}\right)\left(\frac{\Gamma_{AB}\Gamma_\gamma}{\Gamma\hbar}\right)\left(\frac{2I_2^* + 1}{2I_0 + 1}\right) \qquad \text{II-20}$$

From equation II-4, the logarithmic exponents are:

$$n = \frac{E^*}{kT}; \quad m = 0 \qquad \text{II-21}$$

For instance, if $E^* = 6$ MeV, the mean life of a nucleus is $\simeq$

$T_6 = 100 \qquad 1000 \qquad 2000$
$t_{\text{sec}} = 10^{235} \qquad 10^{10} \qquad 10^2$

hence there is an *extremely rapid* variation with temperature.

II–E: *Nuclear reactions in equilibrium with the photon gas*

We now consider the two channels (supposed to be resonant)

$$A + B \rightleftharpoons C^* \rightleftharpoons C + \gamma$$

In equilibrium, we have:

$$\frac{dn_C}{dt} = 0 = n_A n_B \langle\sigma v\rangle_{\text{capt}} - n_C P_{\text{photo}} \qquad \text{II-22}$$

whence, from the equations II-20 and II-14,

$$\frac{n_A n_B}{n_C} = \frac{P_{\text{photo}}}{\langle\sigma v\rangle_{\text{capt}}} = \left(\frac{e^{-Q/kT}}{\lambda_T^3}\right)\frac{(2J_0 + 1)(2S_0 + 1)}{2I_0 + 1} \qquad \text{II-23}$$

where the $\Gamma$ (and hence the reaction rates) do not appear, but where the mass difference $Q$ (and hence the stability of the nuclei) plays a primordial role. This phenomenon will be responsible for the iron peak. We will come back to it later. This result can also be obtained by means of a statistical calculation. We simply consider the final products $A + B \rightleftharpoons C$, and make the hypothesis that equilibrium is reached ($t \gg t_{\text{capt}} + t_{\text{photo}}$). The law of mass action tells us that the abundances are related to the partition functions ($K$) by the equation

$$\frac{n_A n_B}{n_C} = \frac{K_A K_B}{K_C}; \quad K_j = \exp\left(-\frac{Q}{kT}\right)\left(\frac{w_j}{\lambda_{T_j}^3}\right) \qquad \text{II-24}$$

where $Q_j$ is the mass of the particle $j$; $w_j = 2I_j + 1$ (the statistical weight of its ground state) and $\lambda_{T_j} = h/(2\pi m_j kT)^{1/2}$. We deduce:

$$\frac{n_A n_B}{n_C} = \exp\left\{-\frac{(Q_A + Q_B - Q_C)}{kT}\right\}\left(\frac{w_A w_B}{w_C}\right)\left(\frac{\lambda_{T_C}}{\lambda_{T_A}\lambda_{T_B}}\right)^3 \qquad \text{II-25}$$

or,

$$\frac{n_A n_B}{n_C} = \frac{e^{-Q/kT}(2J_0 + 1)(2S_0 + 1)}{\lambda_T^3(2I_0 + 1)}$$

As an example, let us consider the reaction

$$3\text{He}^4 \rightleftharpoons \text{Be}^8 + \text{He}^4 \rightleftharpoons \text{C}^{12*} \to \text{C}^{12}$$

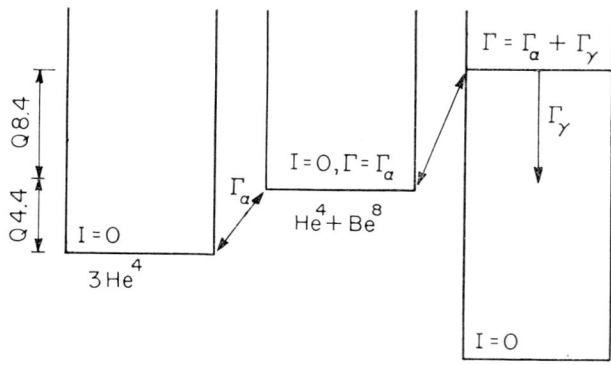

responsible for the luminosity of the red giants.

The first two stages reach equilibrium, the third does not. We first calculate the density of excited carbons in the gas ($n_{12}^*$):

$$\frac{n_{12}^*}{n_4^3} = 3^{3/2}\lambda_{T_4}^6 \, e^{-Q/kT} \qquad \text{II-26}$$

with $Q = Q_{4,4} + Q_{4,8} = 375$ keV.

The rate of $\text{C}^{12}$ synthesis is then given by

$$\frac{dn_{12}}{dt} = n_{12}^* \frac{\Gamma_\gamma}{\hbar} \qquad \text{II-27}$$

## II-F: *The electronic screen*

Until now we have neglected the effect of electrons on the rate of capture.

In the absence of electrons, the classical turning points are located at a distance $R_c$ given by $E = U$, where $U = Z_1 Z_2 e^2/r$, hence $R_c = Z_1 Z_2 e^2/E$.

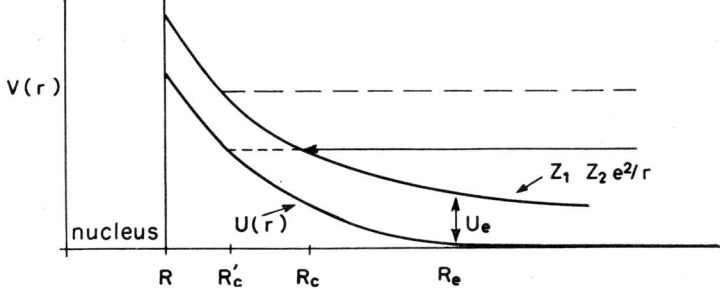

The effect of the electrons is to decrease the electrostatic repulsion, hence to decrease the effective potential.

$$U(r) = \frac{Z_1 Z_2 e^2}{r} - U_e \qquad \text{II-28}$$

In first approximation, we suppose that all the electric charge is concentrated on a layer of radius $R_c$, where $R_e \gg R_c$. We then have:

for $r > R_e$: 
$$U(r) = 0 \quad U_e = \frac{Z_1 Z_2 e^2}{r} \qquad \text{II-29}$$

for $r \alpha R_e$: 
$$U_e = \frac{Z_1 Z_2 e^2}{R_e}$$

The new turning point $R_c'$ is given by

$$E = U = Z_1 Z_2 e^2 \left( \frac{1}{R_c} - \frac{1}{R_e} \right); \quad R_c' = \frac{Z_1 Z_2 e^2}{E + U_e(r = R_e)} \qquad \text{II-30}$$

Whence $R_c' < R_c$. In other words, everything happens as if the incident particle had obtained an energy $U_e(r = R)$ which allows it to penetrate up to $R_c'$; the particles with energy $[E - U_e(r = R_e)]$ now play the role which was previously played by the particles with energy $E$.

We must hence write

$$\langle \sigma v \rangle_{\text{with screening}} = \frac{\int \sigma(E) \, n\{E - U_e(r = R)\} v \, dE}{\int n(E) \, dE}$$

$$= \exp \left( \frac{U_e(r = R_e)}{kT} \right) \langle \sigma v \rangle_{\text{without screening}} \qquad \text{II-31}$$

The rate increase is represented by the multiplicative factor

$$\exp \left\{ \frac{U_e(r = R_e)}{kT} \right\}$$

A more exact calculation allows us to calculate the term $U_e(r = R_e)$. This term becomes large at high densities.

## II-G: *Stellar models*

In the preceding section, we have studied the methods of calculation of thermonuclear reactions from experimental data of nuclear physics. These reactions are of interest for two reasons: they produce energy and influence the course of the stellar evolution; they generate new isotopes and alter the chemical composition of the galaxy.

The second stage of the construction of the mapping function is the building up of stellar models. On top of the generation of nuclear energy, these models introduce other microscopic phenomena which we will not discuss here, such as the atomic phenomena which govern opacity ($\kappa$) and the equation of state of stellar gases.

We first assume that the star is in hydrostatic equilibrium, hence (Lecture I):

$$\frac{dP}{dr} = \frac{-\rho G M_r}{r^2} \qquad \text{II-32}$$

where
$$M_r = \int_0^r 4\pi r^2 \rho \, dr \qquad \text{II-33}$$

Conservation of energy requires that $L_r$, the amount of photons which leaves a sphere of radius $r$, be such that (Lecture I):

$$\frac{dL_r}{dM_r} = \varepsilon_{ph} = \varepsilon_N + \varepsilon_g - \varepsilon_\nu \qquad \text{II-34}$$

Moreover, the luminosity governs the temperature gradient by

$$\frac{dT}{dr} = -\left(\frac{3}{4ac}\right)\left(\frac{\kappa\rho}{T^3}\right)\frac{L_r}{4\pi r^2} \qquad \text{II-35}$$

if the star is in radiative equilibrium. Otherwise,

$$\frac{dT}{dr} = \frac{2}{5}\left(\frac{T}{\rho}\right)\frac{dP}{dr} \quad \text{(convective equilibrium)} \qquad \text{II-36}$$

The solution of these equations for given initial conditions (mass, chemical composition) constitutes a stellar model. Because of the gravitational contraction and the exhaustion of the fuels, the models change with time.

The detailed study of the solution of these equations is a complicated science, with which we will not deal. We will merely state some generalities and describe the results.

## II–H: *The Hertzsprung–Russell diagram*

The diagram carries stellar luminosity ($L$) in the ordinate and surface temperature $T_e$ (or colour) in the abscissa. Each star is represented by a point. By the relation $L/4\pi R^2 = acT_e^4$ (which, in fact, defines $T_e$) it can be seen that stars with small radii are all located in the left bottom corner (dwarfs), and that stars with large radii are in the right upper corner. (The lines $\log(R/R_\odot) = \text{const.}$ are indicated in the diagram.)

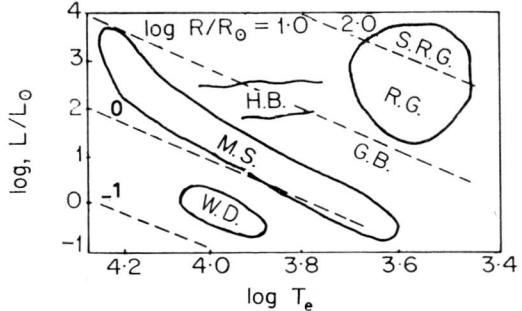

R.S.G. = Red supergiants
R.G. = Red giants
H.B. = Horizontal branch
G.B. = Giant branch
M.S. = Main sequence
W.D. = White dwarfs

During the evolution of a star, the representative point is moving and follows a path. The position of this path depends very much on the mass of the star and to a lesser extent on its chemical composition.

The H–R diagram of stars in the neighbourhood of the sun (cf. diagram) shows well-defined concentrations. Most of the points lie on a diagonal called the Main Sequence. In the upper right corner and in the lower left corner there are two other large concentrations, the red giants and the white dwarfs. Statistically, the concentration of points signifies a slowing down of the motion along the path, and hence a slower evolution of the macroscopic parameters ($L$, $T_e$ or $R$). The earlier discussions suggest a connection between this slowing down and the activity of the nuclear reactions.

We now give some general results of calculations on stellar models.

(a) The stars with homogeneous chemical composition are located on, or to the left, of the M.S.;

(b) The pure-hydrogen models are located on the M.S.;

(c) The models with pure chemical elements heavier than hydrogen are located in layers more or less parallel to the M.S., and at distances which increase with the mean charge (cf. diagram);

(d) Homogeneous models composed of hydrogen and helium (obtained for instance by partial burning of the hydrogen accompanied by

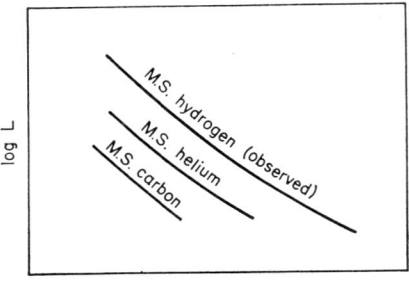

a mechanism of complete mixing of the stellar matter) are located in the region between the M.S. H and the M.S. He, at a *distance proportional to the relative concentrations*.

The absence of helium main sequence and carbon main sequence in the observed diagram excludes the existence of a large amount of stars

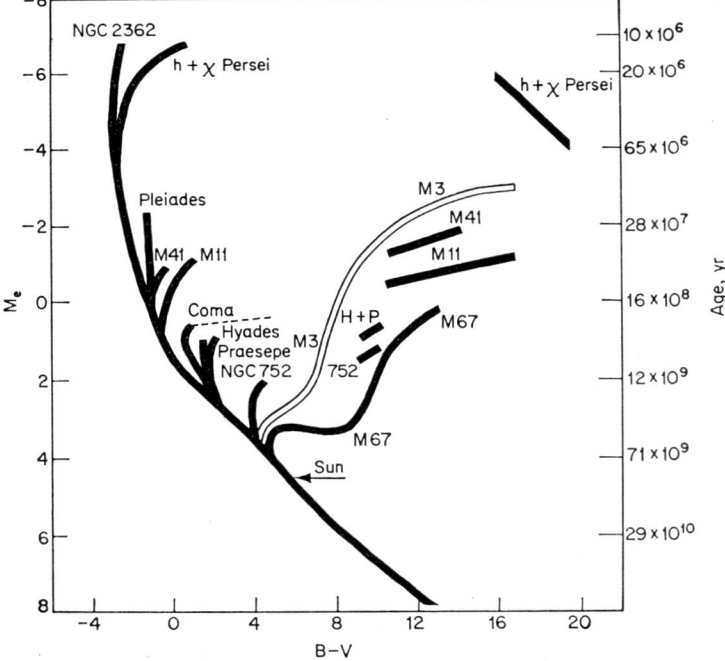

Fig. II–4. Sketch of the H–R diagrams of some star clusters (Sandage, 1957). The ordinate scale is proportional to the total luminosity $L$, and the abscissa to the surface temperature $T_e$. The envelope curve on the left-hand side represents the main sequence. The right-hand parts of M 3, M 41, M 11, M 67 represent red giants (R.G.), whereas those of $h$ and $\chi$ Persei represent red supergiants (R.S.G.). The triangular gap between the M.S. and the group of the R.G. and the R.S.G. is the Hertzsprung gap. The time scale will be discussed in Section II–J.

whose initial composition is completely void of H. The presence of red (hence non-homogeneous) giants illustrates both the alteration by the nuclear reactions of the central chemical composition with respect to the surface composition and the absence of a generalized mixing mechanism in the interior of stars.

It is difficult to analyse further the H–R diagram of the stars in the neighbourhood of the sun; all the masses, the chemical compositions and the ages are possibly represented in it. On the other hand, the H–R of a cluster of stars sharply localized in the sky presents a much simpler problem. Each star probably comes from the condensation of the same initial gas. The initial chemical compositions and the ages are probably the same. Only the masses vary. In Fig. II–4, the H.R. of some star clusters are represented. We must explain them in terms of the theoretical paths mentioned earlier.

II–I: *The H–R paths*

The initial stellar gas is too cold to appear in the H–R. During a short time (several million years) the star contracts. It goes through a

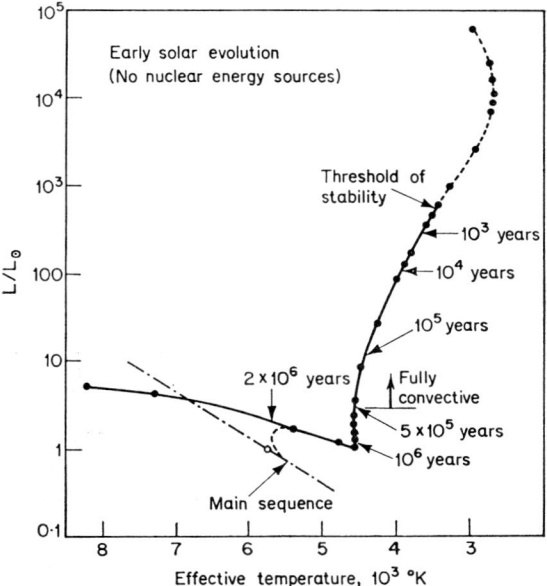

Fig. II–5. The H–R path of the sun before it reached the main sequence (Cameron and Ezer, 1964). In the neighbourhood of $t = 2 \times 10^6$ years, the nuclear reactions (which were not considered in the model) stabilize the star on the M.S.

completely convective stage, then the convective zone moves out towards the surface. It reaches the main sequence when the nuclear reaction $p + p \rightarrow D + e^+ + \nu$ becomes considerable. The path is described in Fig. II–5.

The motion along the path stops momentarily; the concentration of the points generates the main sequence. Stars with very small mass (Fig. II–6) never reach the temperatures necessary for hydrogen burning, they go off directly into the white-dwarf zone. There also the

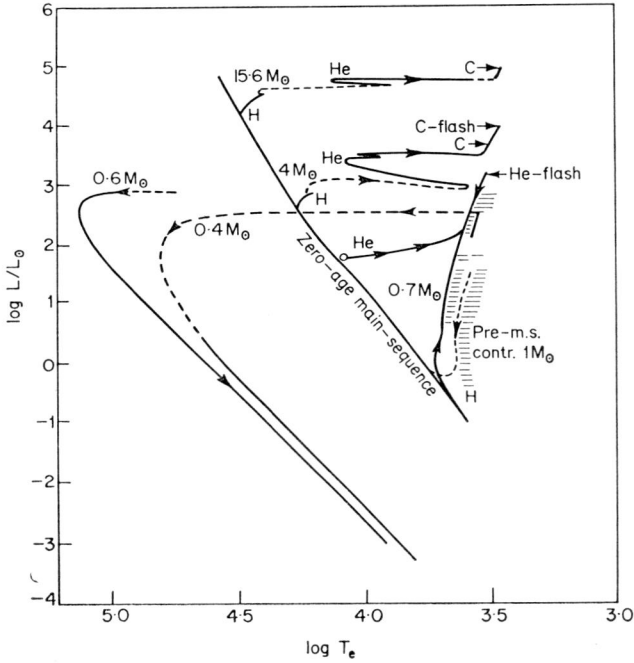

Fig. II–6. H–R paths (Hayashi, 1962) for stars with various masses. We see the burning zones of the H, He and C and the final (still hypothetical) courses towards the domain of the white dwarfs.

path is very much slowed down; the electronic gas reaches the threshold of degeneracy. It is the pressure of the degenerate gas, much greater than that of a perfect gas, which opposes the contraction. Evolution is nearly stopped; result: the group of white dwarfs.

For the stars on the main sequence, the transformation of hydrogen into helium occurs by means of four different mechanisms. They are illustrated in the table (pp chains) and in Fig. II–7. Each of these mechanisms contributes to the emission of energy. ppI dominates for $T_6 \leqslant 8$, ppII for $T_6 < 20$ and ppIV for $T_6 > 30$. ppIII is never important as regards energy; it is of interest because it gives rise to the emission of a neutrino with relatively high energy which one hopes to detect soon in the solar radiation.

The ppIV cycle has another important effect; it transforms all the CNO of the original gas into $\simeq 95\%$ $N^{14}$, $\simeq 1\%$ $C^{13}$, $\simeq 4\%$ $C^{12}$. We will come back to this later.

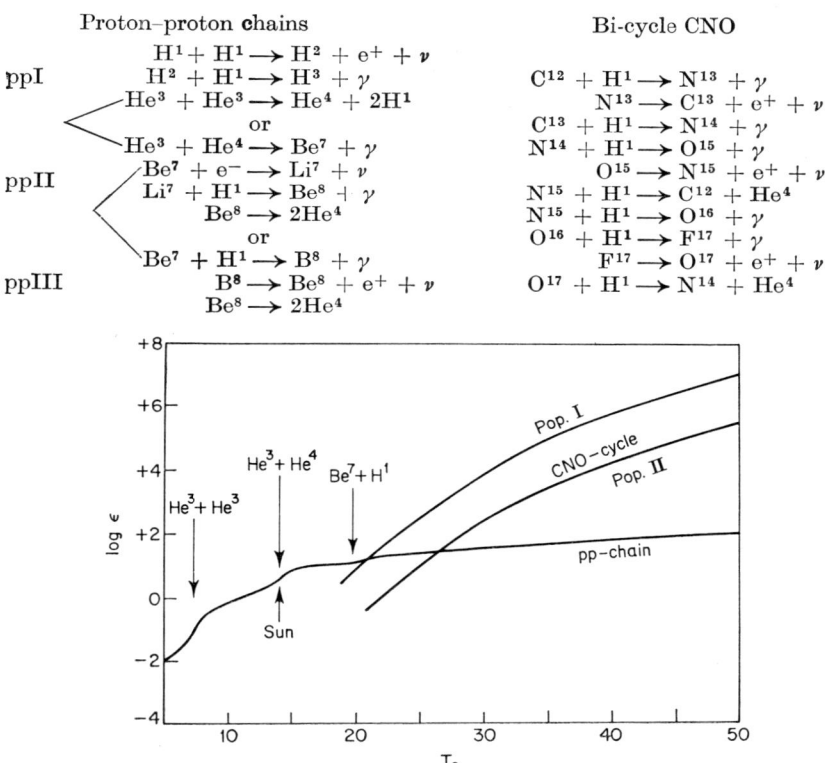

FIG. II–7. Rate of generation of energy (erg/gm/sec) of the transformation $4H \rightarrow He$ at $\rho = 10^2$ gm/cm$^3$ (Fowler, 1959). We indicated the dominant pp branching (see the table). The magnitude of the CNO cycle depends on the initial abundance of CNO, hence on the age of the star. We have taken $X_{\mathrm{CNO}} = 1 \cdot 6 \times 10^{-2}$ for young stars (Pop. I) and $6 \cdot 4 \times 10^{-4}$ for old stars (Pop. II).

As we ascend the main sequence, we encounter stars with *greater and greater mass* and which are *hotter and hotter* in the interior as well as on the surface. The hot stars get their supply of energy from the ppIV. They have a convective nucleus and a radiative envelope. The others live mainly on the 3pp, have a radiative core and a convective envelope.

## II–J: *Duration of the "Main Sequence" phase*

How long will a star remain on the M.S.? Until, because of the production of helium at the centre, the star reaches a degree of inhomogeneity "sufficient" to displace it appreciably towards the region of the red giants. "Sufficient" means in practice that roughly 10% of the initial hydrogen has gone.

The mean value of the concentration of hydrogen (in mass fraction $\bar{X}_1$) of the hydrogen is given by

$$\bar{X}_1 = \int_0^M X_1 \, dM_r \qquad \text{II-37}$$

and luminosity is related to this concentration by

$$L = -\frac{d}{dt}(M\bar{X}_1)Q_{pp}, \quad Q_{pp} = 6\cdot 7 \text{ MeV/nucleon} \qquad \text{II-38}$$

whence the fusion time of the hydrogen:

$$t_H = -\int_{\bar{X}_{1\,\text{initial}}}^{\bar{X}_1(t)} Q_{pp}\left(\frac{M}{L}\right) dX_1 = \left(\frac{M}{L}\right)(\bar{X}_{1\,\text{initial}} - \bar{X}_1(t))Q_{pp} \qquad \text{II-39}$$

For the sun $L/M \simeq 2$ erg/gm/sec, so that $t \simeq 10^{10}$ years. For an arbitrary star on the M.S.

$$t_H = t_{M.S.} = \left(\frac{M}{M_\odot}\right)\left(\frac{L_\odot}{L}\right) \times 10^{10} \text{ years} \qquad \text{II-40}$$

## II–K: *The mass–luminosity relation*

A dimensional analysis allows us to evaluate very approximately the ratio $L/M$ in terms of $M$. First, the radiative transfer equation

$$L_r = -\left(\frac{4\pi r^2}{3\kappa\rho}\right)\left(\frac{d}{dr}\right)(acT^4)$$

when roughly integrated, gives us $L \propto R^4 T^4/M$.

The pressure of gravitational origin is then given by $\bar{P} = M^2 R^4$ (from $dP/dr = -\rho G M_r/r^2$). If the weight of the star is maintained by the pressure of the gas (then non-degenerate), $P \propto \rho T \propto M^2/R^4$, whence $T \propto M/R$ and $L \propto M^3$. This is the case for stars $0\cdot 5 < M < 10 M_\odot$. If the weight of the star is maintained by the radiation pressure $P \propto T^4 \propto M^2/R^4$, then $L \propto M$. This is the case for stars with greater mass.

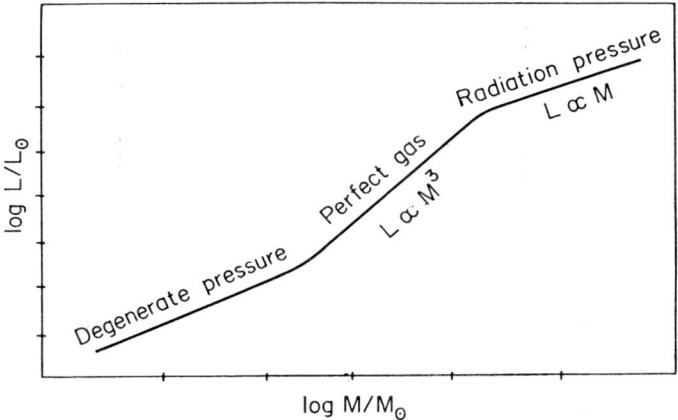

We get thus, approximately,

$$t_{M.S.} \simeq 10^{10} \text{ years}/M^2 \quad 0.5 < M < 10 M_\odot \qquad \text{II-41}$$
$$t_{M.S.} \simeq 10^7 \text{ years} \qquad M \gtrsim 20 M_\odot$$

Detailed calculation shows that the limit period is $\simeq 3 \times 10^6$ years (see Fig. II-8).

We see thus that the evolution of a stellar cluster containing stars of the same age and same chemical composition, but of various masses, can be described in the following manner. After the phase of gravitational contraction (which is very short) the stars end up on the M.S.: this is called the zero-age. Then the stars with greater mass (hence at the higher end of the M.S.) move to the right. The exodus continues gradually, involving stars located lower and lower on the M.S. The evolution of a cluster after several times $t$ is described in Fig. II-9. This behaviour is well illustrated experimentally if we superimpose on the same diagram the star distributions of several clusters (Fig. II-4).

We see thus how the age of a cluster can be measured. We determine the luminosity and the corresponding mass at the break-off point in the H–R of a given cluster and we apply relation (II-39).

## II–L: *The red-giant phase*

Why do inhomogeneous stars migrate to the right? We now give a partial qualitative explanation of this fact. For a perfect gas $P = nkT = (\rho/\mu)(kT/m_H)$ where $n$ is the number of particles per unit

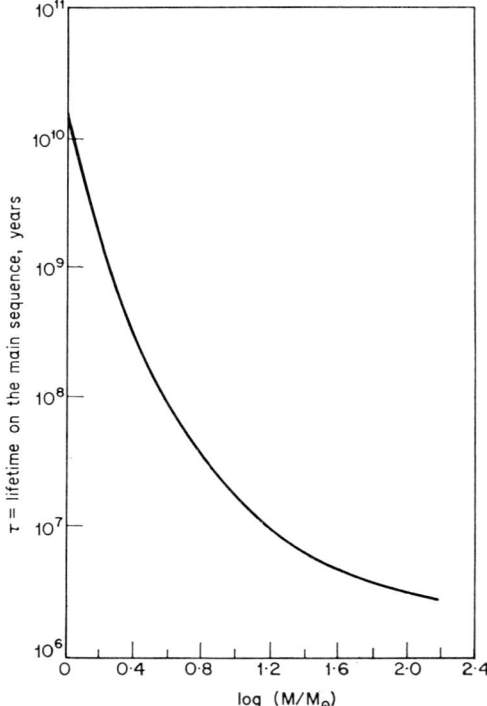

Fig. II-8. Lifetime on the Main Sequence as a function of Stellar Mass (Limber.)

volume and $1/\mu = (Z + 1)/A$, the number of particles per nucleon. For hydrogen $1/\mu = 2$ and for helium $1/\mu = \frac{3}{4}$. Thermal and mechanical equilibrium requires that $P$ and $T$, and hence also $n$, be continuous in the whole star. Since at the interface between a hydrogen zone and a helium zone, $\mu$ has to undergo a sudden change, $\rho$ will have to do the same.

In the diagram, the variation of $\rho$ and of $\mu$ in terms of the central increase in helium is illustrated *very schematically*.

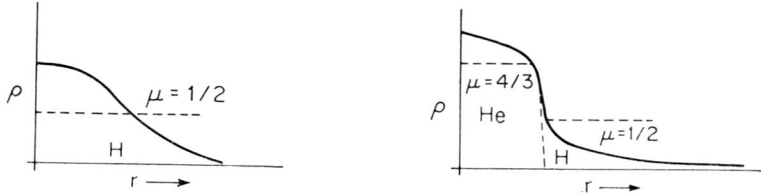

Everything happens as if the envelope were pushed back to the exterior. The radius increases; the star moves towards the region of the giants.

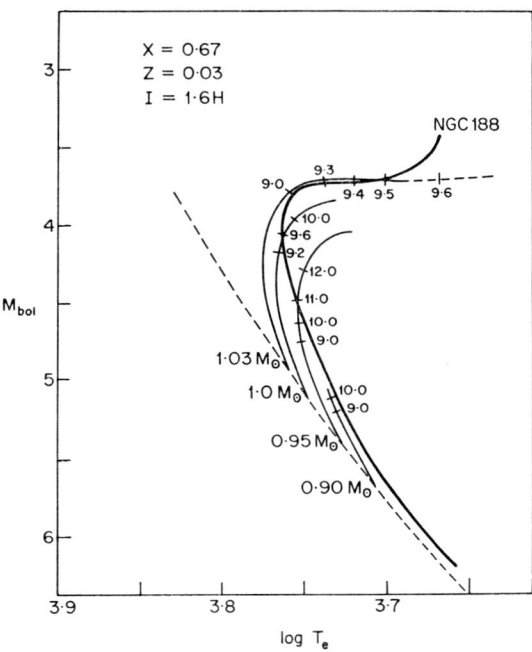

FIG. II-9. Each path represents the evolution of a star with given mass after its stage on the main sequence (Demarque, 1964). The numbers on the paths represent the ages of the stars (in $10^9$ years) when they reach the corresponding dashes. The full line of the NGC188 cluster crosses these paths at $t \simeq 9 \times 10^{10}$ years.

II–M: *Shell sources*

The star begins to contract again; it heats up. The layer of hydrogen contiguous to the helium zone takes fire. It is the CNO cycle which prevails. Here the stars with non-degenerate nucleus ($M \geqslant 2M_\odot$) evolve very rapidly, whereas the lighter stars see their pressure increase considerably because of the degeneracy.

The contraction continues until new nuclear reactions occur. We will first have $N^{14}$ (prepared by the CNO cycle), then, around $T_6 = 100$, the reaction $3He^4 \rightarrow C^{12}$. The path slows down again. In the H–R diagram, the phase of gravitational contraction gives rise to the Hertzsprung gap (Fig. II–4), an empty region (hence very rapidly traversed) of triangular shape. At the lower basis of the triangle, the degenerate pressure already plays its moderating role.

## II-N: *Flash*

In a medium where the electrons are strongly degenerate, the pressure and the temperature are nearly completely uncoupled. The equation of state takes the form

$$P \propto \rho^{5/3}\left\{1 + a\left(\frac{kT}{E_F}\right)^2 + \cdots\right\} \qquad \text{II-42}$$

At this time we have $kT \simeq 10$ keV and $E_R \simeq 100$ keV (the Fermi energy); $a$ is a constant of the order of unity. When the temperature will be such that the nuclear reactions can begin, the pressure will hardly notice the difference. The hydrostatic readjustment characteristic of perfect gases will not be able to occur. The temperature will zoom up, at the same time accelerating the rate of the nuclear reactions, and vice versa. Everything will settle down when, owing to the increase in temperature, the degeneracy will have disappeared.

Since this "explosion" only lasts a very short time, it will be hardly visible from outside. However, this phenomenon has often been put forward as the explanation of the ejection of matter by the planetary nebulae (shock wave propagating to the surface). After the flash, the star continues, though it has cooled down, to burn its helium into carbon. In the H–R diagram, the star lies at the extremity of the red-giant branch (Fig. II-4).

## II-O: *The helium burning*

In Section II-E we described the mode of production of energy by helium burning. Since the nuclei of mass 5 and 8 are all unstable, we must wait for the triple reaction $3\text{He}^4 \rightarrow \text{C}^{12}$ in order to obtain a suitable source of energy. In fact, it is a reaction in two stages, one of which rests on the "weak" instability of the $\text{Be}^8$ (+96 keV). A certain concentration of $\text{Be}^8$ settles in, resulting from the reactions of capture and dissociation ($2\text{He}^4 \rightleftharpoons \text{Be}^8$), then we have ($\text{Be}^8 + \text{He}^4 \rightleftharpoons \text{C}^{12*} \rightarrow \gamma + \text{C}^{12}$).

The effective exponent of this reaction is very high

$$(n = \{Q/(kT) - 3\} \simeq 30)$$

where $Q$ is the mass difference between $3\text{He}^4$ and $\text{C}^{12*}$ (375 keV). Such an exponent creates a temperature gradient which makes the stellar nucleus convective.

The stellar core becomes richer in $C^{12}$. We will soon have to take the eventual sequence

$$C^{12} + He^4 \rightarrow O^{16} + \gamma$$
$$O^{16} + He^4 \rightarrow Ne^{20} + \gamma$$
$$Ne^{20} + He^4 \rightarrow Mg^{24} + \gamma$$

into account.

All these targets have zero spin, as does $He^4$. The only possible values for the spin of the capture levels are thus $0^+$, $1^-$, $2^+$, etc. The situation is described in Figs. II–I, II–10, and II–11. The vertical lines locate the Gamow peaks in terms of the temperature.

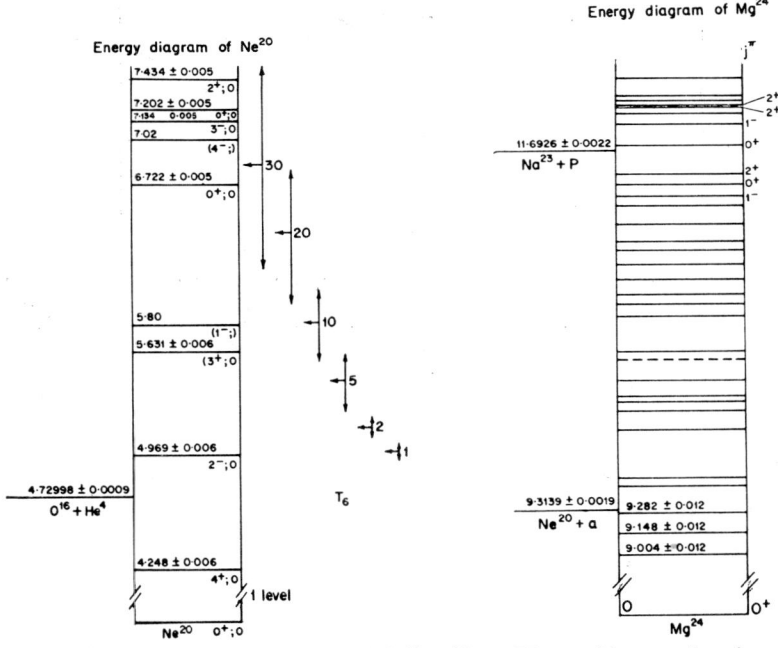

FIG. II–10. Energy diagram of $Ne^{20}$. On the right-hand side, the position and the width of the Gamow peaks for various temperatures ($T_8 = 10^8$ °K). Only the levels with natural parity ($0^+$, $1^-$, $2^+$, $3^-$, etc.) can be formed by the collision $He^4 + O^{16}$.

FIG. II–11. Energy levels of $Mg^{24}$. Unlike the cases $C^{12} + He^4$ and $O^{16} + He^4$, the reaction $Ne^{20} + He^4$ corresponds to a rather thickly populated region of the energy diagram.

The reaction $C^{12} + He^4$ occurs on the resonance wing at 7·118 MeV in $O^{16}$. The rate of reaction is very little known (up to a factor 10). The other rates are better known.

Comparing the three diagrams, we notice the difference in density of levels between the first two and the last in the neighbourhood of the $(\alpha, \gamma)$ capture threshold. From this there follows a huge difference in reaction rates. Consequently, $Ne^{20}$, which is formed slowly and rapidly destroyed, is never an important product of the helium-burning phase. The situation is illustrated in Figs. II–12 and II–13.

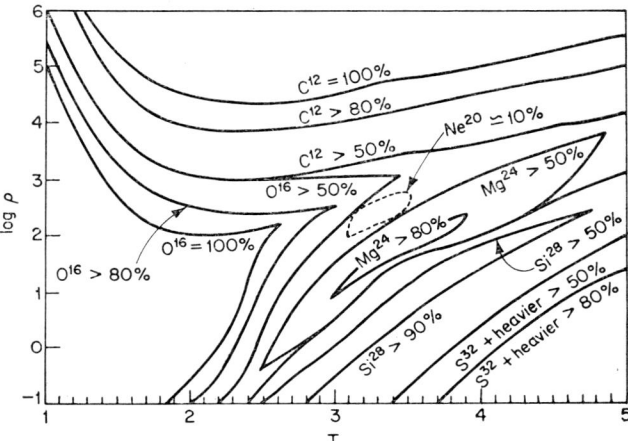

Fig. II–12. The isotopic products of helium burning for fixed temperature and density are described by means of an iso-abundance diagram. For instance, the caption $O^{16} > 50\%$ in a region means that in this region oxygen forms more than 50% of the total mass. $Ne^{20}$ is never dominant. It reaches at most 10% in a small region.

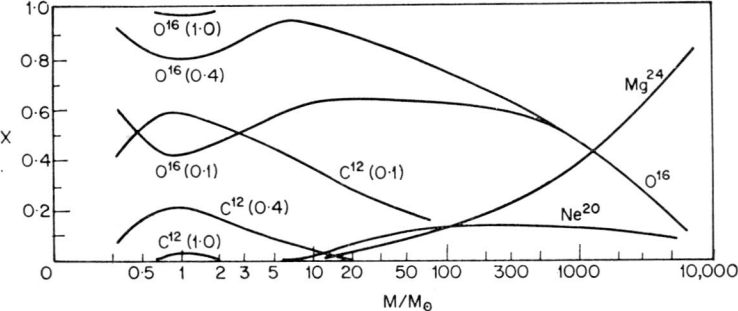

Fig. II–13. Isotopic abundances at the end of the helium burning in terms of the stellar mass (Deinzer and Salpeter, 1964.) The three curves for $C^{12}$ and $O^{16}$ represent the choices of 0·1, 0·4 and 1·0 for the values of the parameter $\theta_\alpha^2$ of the reaction $C^{12} + He^4 \rightarrow O^{16} + \gamma$. The values of $Ne^{20} + Mg^{24}$ are practically independent of the choice of $\theta_\alpha^2$. The only stars which can produce $Ne^{20}$ and $Mg^{24}$ have mass $M > 50 M_\odot$. Such stars are very rare; they cannot be playing an important nucleosynthetic role. It is to the carbon burning that we will attribute these elements. Note: Recent experiments have shown $\theta_\alpha^2 \simeq 0 \cdot 1$.

In almost all realistic circumstances, mostly $C^{12}$ and $O^{16}$ are formed, in relatively comparable amounts, where the exact ratio depends on the little-known rate of the reaction $C^{12} + He^4 \rightarrow O^{16} + \gamma$. In very large stars, $Mg^{24}$ and $Si^{28}$ are perhaps formed also. From the viewpoint of the nucleosynthesis, we must mention the sequence

$$N^{14}(\alpha, \gamma)\ F^{18}(\beta, \nu)\ O^{18}(\alpha, \gamma)\ Ne^{22}(\alpha, \gamma)\ Mg^{26}$$
$$\text{or } (\alpha, n)\ Mg^{25}$$

as a source of neutrons. We will come back to it later.

Stars with $M < 0.5 M_\odot$ do not burn the helium; they now emigrate to the white-dwarf zone.

The corresponding H–R paths during the burning phase of the helium are rather complex. We will illustrate them by means of two examples: $M = 3M_\odot$ (Iben) and $M = 7M_\odot$ (Kippenhahn et al.). They appear in Figs. II–14 and II–15.

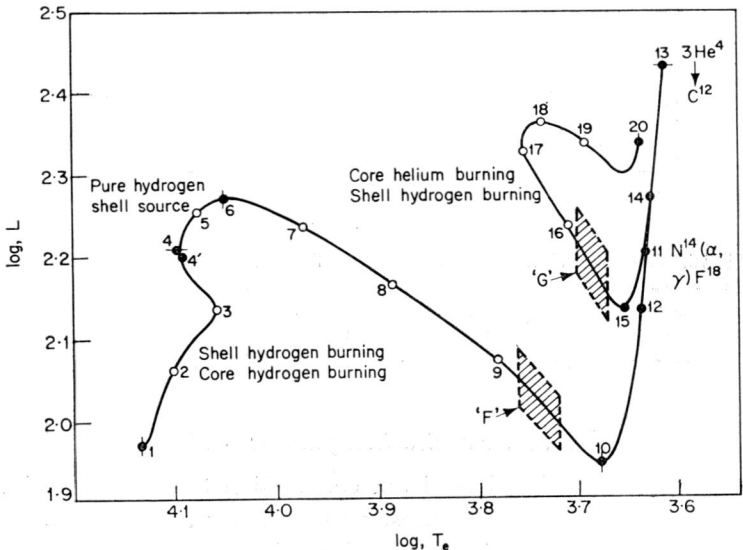

Fig. II–14. H–R path of a star of $3M_\odot$ during the helium burning (Iben, 1964).

The path of the $7M_\odot$ star crosses five times the region in which the famous Cepheid variables are located. Now during each of these crossings, the star becomes unstable and liable to pulsate radially. This might be an acceptable model of Cepheid stars. Fig. II–16 describes the internal constitution of a star with $M = 7M_\odot$, as a function of its age.

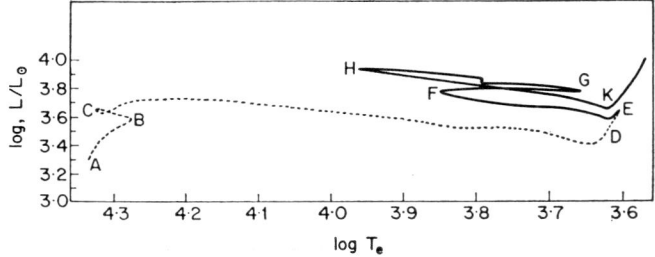

Fig. II–15. H–R path of a star of $7M_\odot$ during the helium burning (Kippenhahn, 1964).

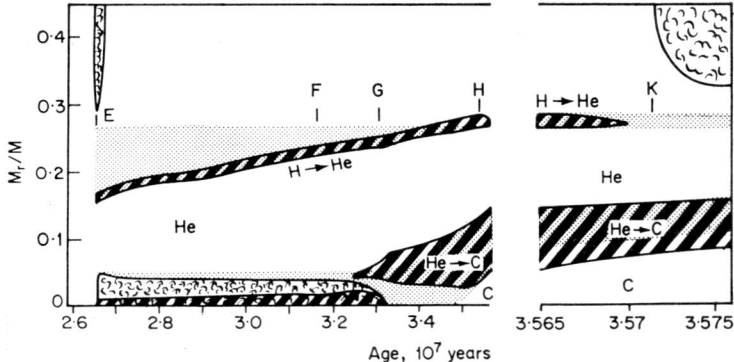

Fig. II–16. The evolution of the structure of the $7M_\odot$ star (Fig. II–15). In the ordinate we have the mass ratio, in the abscissa, the age in $10^7$ years. The hatched zones are zones of nuclear burning, the streaked zones are convective, and the dotted zones are zones which are enriched by the products of the burning. The letters correspond to the preceding diagram (Kippenhahn, 1964).

II–P: *New contraction phases: neutrinos*

When the helium is all used up, contraction takes over again. The temperature rises again. We must now consider the emission of neutrinos, first by the excitation mechanism of the electronic plasma (high density) or the photoneutrino mechanism, or the mechanism of pair annihilation (cf. Lecture III).

Detailed models are much rarer here. Kippenhahn *et al.* (1965) show that even a $5M_\odot$ star is now strongly affected by the degeneracy and that it will undergo a period of cooling down at the centre; the energy comes from a layer of H which is nearly at the surface. The central density increases unceasingly; will we get an enormous flash?

II–Q: *The carbon burning*

The carbon originating from the preceding phase will become a nuclear fuel at $T_6 = 700$–$900$. The four-branch reaction

$$\begin{aligned}
C^{12} + C^{12} &\to Ne^{20} + He^4 & 50\% & \quad Q = 4\cdot 616 \text{ MeV} \\
&\to Na^{23} + H & 50\% & \quad Q = 2\cdot 237 \text{ MeV} \\
&\to Mg^{24} + \gamma & \text{very rare} & \quad Q = 13\cdot 930 \text{ MeV} \\
&\to Mg^{23} + n & \text{rare} & \quad Q = -2\cdot 602 \text{ MeV}
\end{aligned}$$

The generated $He^4$ and $H$ will in turn react and produce the isotopes $Mg^{24,25,26}$, $Al^{27}$, $Si^{28}$, in decreasing amount. Quantitatively, the net result is $Ne^{20} \simeq 30\%$, $Na^{23} \simeq 10\%$, $Mg \simeq 60\%$.

The last branch represents a source of neutrons which is perhaps of importance. We will come back to it later.

There are no detailed models here; only simplified models or purely energy models.

The energy model is based on the hypothesis that the photon emission becomes at this stage much lower than the neutrino emission. The carbon-burning period is then characterized by the equation

$$\varepsilon^N_{C+C}(T, \rho) = \varepsilon_\nu(T, \rho) \qquad \text{II–43}$$

The values of $\rho$ and $T$ for which this relation holds determine a curve in the $(\rho, T)$ plane (Fig. II–17). Below this curve, $\varepsilon_N < \varepsilon_\nu$. The evolution of the parameters $\rho_c$ and $T_c$ of a given star is such that the representative point comes from the lower left corner and goes towards the curve, where it stops. We can then calculate, for each point of the curve, the

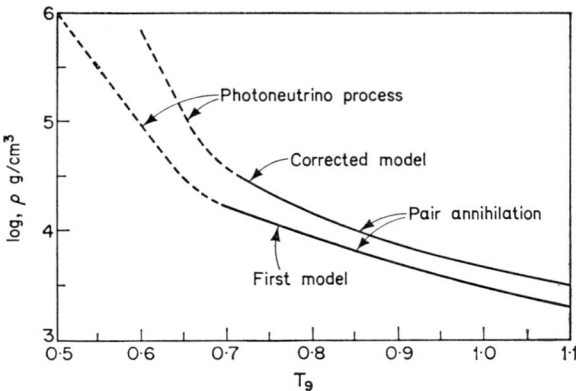

Fig. II–17. Results of the energy model for the carbon-burning stage (Reeves, 1963). The "first model" curve describes the central stellar conditions when the rate of nuclear energy ($C^{12} + C^{12}$) is equal to the rate of neutrino dissipation at the centre of the star. The "corrected model" curve takes certain corrections into account. The dominant dissipative process is identified on the curve.

burning time of the carbon and the emitted energy (Fig. II–18). We can also calculate the corresponding stellar mass at each of the points (by means of a polytropic model) but this evaluation is not very accurate.

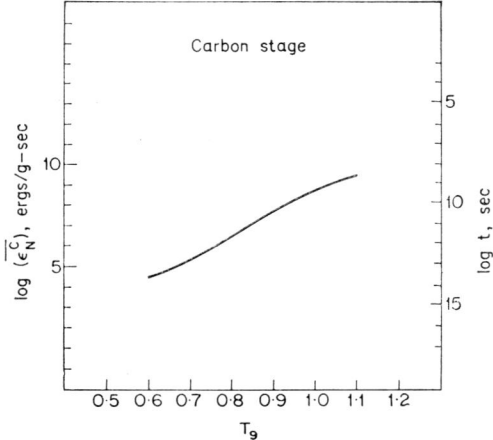

FIG. II–18. From the curve determined in Fig. II–17, we can calculate the rate of energy emission and the lifetime of a star during the carbon-burning stage.

The polytropic models of Hayashi give us some more information on the H–R paths. The stars with $M < 0.7 M_\odot$ never burn their carbon. The stars with large mass ($\simeq 15 M_\odot$) burn their carbon in the red-supergiant region. Unfortunately, Hayashi neglected the effect of the neutrinos on the model. Deinzer and Salpeter, with star models of pure carbon, displayed the importance of the neutrinic phenomena; they

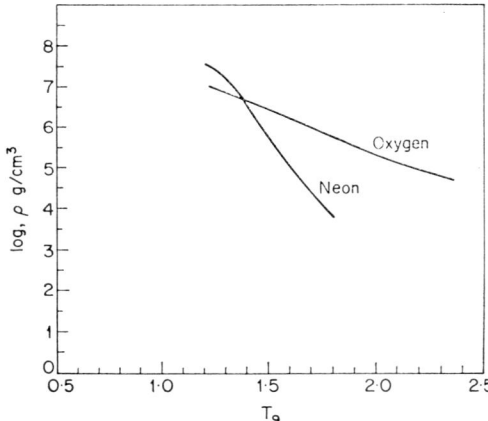

FIG. II–19. Energy model for the neon and oxygen burning stage.

raise the central operating temperature high above the values estimated by Hayashi (but in agreement with the energy models).

### II–R: *Neon photodisintegration*

The contraction phase which follows the exhaustion of the carbon is momentarily perturbed by the sequence $Ne^{20} + \gamma \rightarrow O^{16} + He^4$, followed mostly by $Ne^{20} + \alpha \rightarrow Mg^{24} + \gamma$. Hence, briefly, $2Ne^{20} \rightarrow O^{16} + Mg^{24}$, $Q = 4.58$ MeV, $n \simeq 60$. We do not have here a real stage of nuclear burning: the contraction is, at most, slowed down (Fig. II–19).

### II–S: *The oxygen burning*

In the neighbourhood of $T_6 = 1400$, the $O^{16}$ burns

$$\begin{aligned}
O^{16} + O^{16} &\rightarrow Si^{28} + He^4 & \simeq 45\% \quad & Q = 9.593 \text{ MeV} \\
&\rightarrow P^{31} + H & \simeq 45\% \quad & Q = 7.676 \text{ MeV} \\
&\rightarrow S^{32} + \gamma & \text{very rare} \quad & Q = 16.538 \text{ MeV} \\
&\rightarrow S^{31} + n & \simeq 10\% \quad & Q = 1.459 \text{ MeV}
\end{aligned}$$

According to Hayashi, the corresponding H–R region is still the domain of the red supergiants. According to the energy model, this stage lasts for such a short time that there is little chance of observing such stars (Fig. II–19). Some Si, P and S are produced here and, by the effect of the $p$, $n$ and $\alpha$, probably also some Cl and A.

*Later nuclear reactions*

We might expect a burning stage for the $Mg^{24}$ ($Mg^{24} + Mg^{24}$), for the $Si^{28}$ ($Si^{28} + Si^{28}$). In fact, in the neighbourhood of $T_6 = 2000$, the photodisintegration reactions become so rapid that they play a dominant energy role; the Mg and the Si, etc. disappear before they can generate their stages.

Now, the rates of nuclear interactions will no longer be governed by the charges but by the mass differences. This will be the subject of the next lesson.

### II–T: *The mapping effect*

To certain groups of stars in the H–R plane, we associated certain well-determined nuclear reactions. These nuclear reactions must provide for the star during the corresponding stage. These nuclear reactions

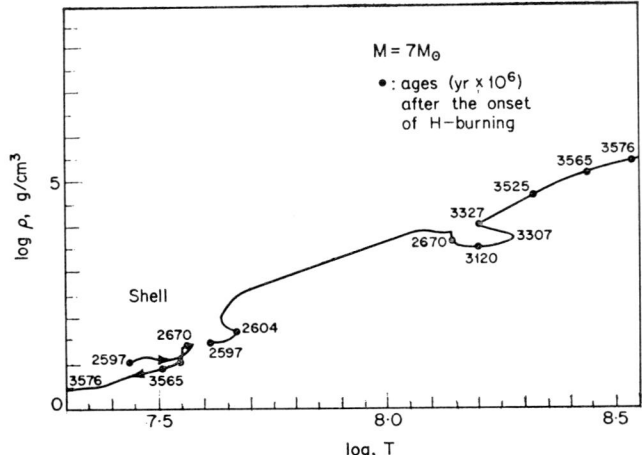

Fig. II-20. Detailed evolution of the centre of a $7M_\odot$ star during the red-giant phase (Kippenhahn). To the left, the hydrogen-burning layer; to the right, the central helium burning.

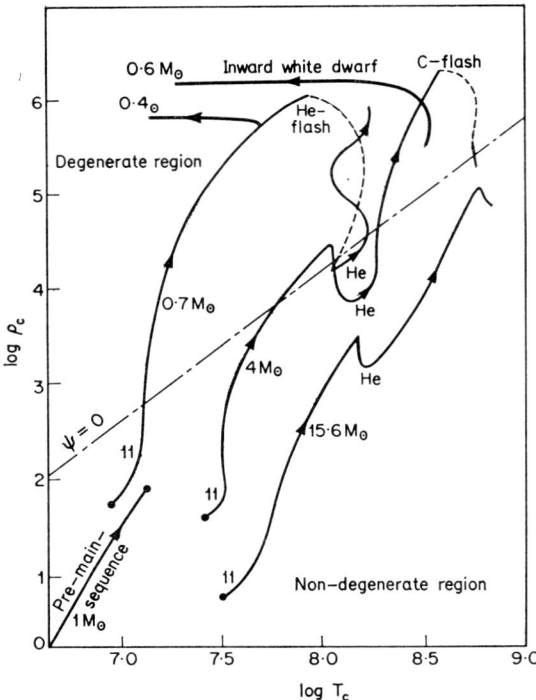

Fig. II-21. Evolution of the central parameters of stars of various masses (Hayashi, 1962).

are, moreover, responsible for the generation of the light isotopes ($A < 40$), if we make exception of the Li, Be, B group. The final isotopic distributions depend on the temperature $T_c$, the density $\rho_c$ and the initial chemical composition $n_i$ at the centre of the star. These parameters vary with time. Certain evolutionary sequences are illustrated in Figs. II–20 and II–21. Using such sequences, we can perform detailed calculations of the nucleosynthetic evolution.

The mapping from the H–R plane to the $Z$–$N$ plane is illustrated in the following diagram.

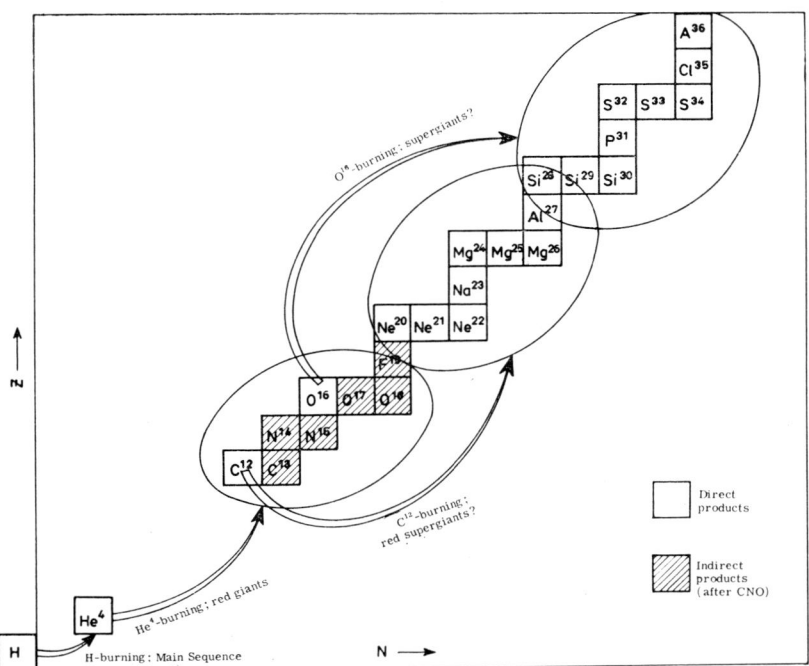

It is up to the main sequence to produce the helium (roughly 20% of the hydrogen in mass). The catalytic reactions implied in the transformation $4H \to He^4$ also have important nucleosynthetic effects on the light isotopes which are already present in the stellar gas; the three P.P. cycles essentially transform all the light isotopes D, $He^3$, $Li^6$, $Li^7$, $Be^9$, $B^{10}$, $B^{11}$ into $He^4$. The CNO cycle generates, from $C^{12}$ and $O^{16}$, the various stable isotopes of C, N, O, F.

It is up to the red-giant branch to transform the helium into $C^{12}$ and

$O^{16}$ (Figs. II–12 and II–13); these isotopes can then be altered by the CNO cycle in a star of later generation.

It is up to the red supergiants, it seems, to transform the $C^{12}$ into isotopes Ne, Na, Mg, Al and, at least partially, $Si^{28}$. Also the $O^{16}$ into Si, P, S, Cl, and perhaps A, K, etc. but in very small amounts.

Other reactions will come and change the abundances of these components, but the general appearance of the curve will remain the same.

All these reactions are controlled by the Coulomb barrier between the charges of the nuclei in question. The more the nuclei are charged, the more the reactions are delayed and the higher the operational temperatures must be. The necessary physical conditions are more difficult to obtain, and hence more rare in nature. Very roughly, this is the explanation of the descending slope of the first peak in the diagram of the elemental abundances, a peak which goes from H to $A \simeq 40$. A detailed theory of the relative abundance of these elements requires knowledge of the stellar statistics and of the restitution mechanisms. We will consider this question in Lecture V.

*Bibliography*

A good start:

J. M. BLATT, V. F. WEISSKOPF, "Theoretical Nuclear Physics", Chap. VIII and IX, John Wiley & Sons (1952)
The article by P. MORRISON in "Experimental Nuclear Physics", (Segrè), Wiley (1961)

Rate of generation of energy:

W. A. FOWLER, J. L. VOGL, "Nuclear and Neutrino Processes in Stars and Supernovae", Lectures in Theoretical Physics, Vol. VI, University of Colorado Press (1963)
H. REEVES, "Stellar Energy Sources" (1963), in Vol. VIII of "Stars and Stellar Systems"
 "A Review of Nuclear Energy Generation in Stars" (1964) (Goddard Institute for Space Studies), in "Stellar Evolution", Plenum Press, New York, 1966

Stellar models:

H. WRUBEL, "Stellar Interiors", *Handbuch der Physik*, Vol. LI (1958)
C. HAYASHI, R. HOSHI, D. SUGIMOTO, *Progr. Theor. Phys.* Suppl. **22** (1962)
P. R. DEMARQUE, R. B. LARSON, "The Age of the Galactic Cluster N.G.C.188", *Ap. J.* **140**, 544, 1964
I. IBEN, "From the Main Sequence through Core Helium Burning", *Ap. J.* **147**, 650, 1967
R. KIPPENHAHN, E. HOFMEISTER, A. WEIGERT, "Die Helium brennende Phase und die Cepherdenstadien eines Sterns von $7 \cdot 0 M_\odot$", *Zeitschrift für Astrophysik*, **60**, 57, 1964

M. Schwarzschild, R. Härm, "Red Giants of Population II", *Ap. J.* **163**, 158, 1962

W. Deinzer, E. E. Salpeter (1965), *Ap. J.* **142**, 813, 1965

"Modèles d'étoiles et Evolution stellaire", Vol. 16 of Congrès et Colloquesed l'Université de Liège

# III

## The role of gammas and neutrinos; the iron peak

DURING the stages of hydrogen and helium burning, the E.M. interaction has played two distinct roles:

(1) transfer of energy outside the star, by means of radiative transfer;
(2) control of the rate of thermonuclear reaction by means of electrostatic repulsion.

Now, with the rise in temperature, the photons will get enough energy as to be able to participate more actively in the stellar evolution and the nucleosynthesis.

### III–A: *Neutrinos: the shadow of a doubt*

When a certain fraction of the photons of the thermal distribution have energies neighbouring $\simeq 0.5$ MeV, the phenomenon of pair creation $\gamma \to e^+ + e^-$ (real or virtual) comes into play to a considerable extent. The pairs created in this way annihilate one another and give us again $e^+ + e^- \to \gamma$.

Some years ago, Pontecorvo pointed out the following fact: the theory of weak interactions, as it is formulated in particular by Feynmann and Gell-Mann, admits the existence of the reaction $e^+ + e^- \to \nu + \bar{\nu}$, that is to say the possibility of a direct coupling between electrons and neutrinos in the absence of nucleons. This statement is based on the form of the Fermi "interaction current". The set of phenomena generated by the weak interaction, and their respective transition amplitudes is proportional to the product

$$JJ^+ \quad \text{where } J = (\bar{n}p) + (\bar{e}\nu) + (\bar{\mu}\nu)$$

($J^+$ is the complex conjugate). The product $JJ^+$ will thus contain a set of nine terms, of which three are "square" terms and six are "cross" terms. For instance, the cross term $(\bar{n}p)(\bar{\nu}e)$, which can be rewritten in the form $n \to p^+ + e^- + \bar{\nu}$, corresponds to the neutron disintegration; the term $(\bar{\mu}\nu)(\bar{p}n) \Rightarrow \bar{\mu} + p \to n + \bar{\nu}$ corresponds to the capture of a muon by a proton, and the term $(\bar{\mu}\nu)(\bar{\nu}e) \to \bar{\mu} \to e^- + \nu + \bar{\nu}$ gives the disintegration of a free muon. Now all these phenomena are observed, and their relative amplitude is indeed given by the Feynmann and Gell-Mann theory. None of the "square" terms has yet been observed. The theory assigns the same relative probability to them as it does to the others, but experiments remain too difficult.

The following objection has been raised: is the prediction of the existence of these cross terms merely formal or does it correspond to reality? In other words, can a theory of weak interactions be stated which takes the observed terms into account, but ignores the others? In this sense, there is still a shadow of a doubt on the reality of the processes under consideration.

On the other hand, if, as is believed, muons and electrons are, up to their mass, identical twins, then the existence of the $\bar{\mu} \to e^- + \bar{\nu} + \nu$ reaction obviously suggests the existence of $e^- \to e^- + \nu + \bar{\nu}$, and consequently of $e^+ + e^- \to \bar{\nu} + \bar{\nu}$. This is a suggestion, not a proof; otherwise we go round in circles.

As we have seen earlier, the action of these neutrinos deeply influences the course of stellar evolution. Inversely, it is possible to compare with reality stellar models which either take these reactions into account or ignore them. Such models can, in certain cases, be profoundly different. It is likely that comparison with observation will allow us to take sides and thus, perhaps, we will be able to find out in the sky what weak interaction does to these square terms.

### III–B: *Neutrinos: simplified calculation*

In stellar conditions, three types of neutrino interactions must be considered. They will be qualitatively described and approximate methods of calculation will be indicated, methods which will allow us to understand the physical nature of the processes in question. Our starting point is the following: For each electronic process which leads to emission of one photon there exists a branch which leads to emission of a pair of neutrinos; the ratio being

$$\frac{P(\nu\bar{\nu})}{P(\gamma)} = \frac{G^2}{(e^2/\hbar c)^4}\left(\frac{E}{mc^2}\right)^4$$

where $G = 1\cdot 0 \times 10^{-5}(M_e/M_p)^2 \simeq 3 \times 10^{-12}$ is the coupling constant of the weak interactions, and $(e^2/\hbar c) = 1/137$ is the electromagnetic interactions constant.

III–B–1) *Photoneutrinos.* We first consider the process

$$\gamma + e^- \rightarrow e^- + \gamma$$
$$\searrow e^- + \nu + \bar{\nu}$$

and make the following approximations: (a) the important photons are such that $kT \ll h\nu \ll mc^2$ (valid for $T_6 \ll 5000$); (b) the electron recoil is neglected (the neutrinos absorb all the energy $h\nu$ of the incident photon). The cross-section for $\gamma + e^- \rightarrow e^- + \gamma$ is proportional to the square of the classical radius of the electron $r_0$: $(r_0 = e^2/mc^2 \simeq 2\cdot 8 \text{ f})$, $\sigma_0 = 8\pi r_0^2/3$ (Thompson scattering). The energy emitted by the photons with energy between $\nu$ and $\nu + d\nu$, per unit volume (where there are $n_e$ electrons) per second, can be written as follows:

$$dE = (8\pi\nu^2 d\nu\ e^{-h\nu/kT})(n_e)\left[(\sigma_0 c)\left\{\begin{matrix}P(\nu\bar{\nu})\\P(\gamma)\end{matrix}\right\}\right](h\nu) \qquad \text{III–1}$$

$$\underbrace{\phantom{(8\pi\nu^2 d\nu\ e^{-h\nu/kT})}}_{\text{no. of photons}} \underbrace{\phantom{(n_e)}}_{\text{no. of }e^-} \underbrace{\phantom{(\sigma_0 c)P(\nu\bar{\nu})/P(\gamma)}}_{\substack{\text{prob. of emis-}\\\text{sion of a }\nu\bar{\nu}}} \underbrace{\phantom{(h\nu)}}_{\substack{\text{emitted}\\\text{energy}}}$$

$$\propto n_e(kT)^8\ e^{-h\nu/kT}d\left(\frac{h\nu}{kT}\right)\left(\frac{h\nu}{kT}\right)^7$$

The integrant has a peak $\Delta E \simeq kT$, for an energy $h\nu \simeq 7kT$. The energy emitted per gram is obtained by means of $\varepsilon_\text{photo} = \int dE/\rho \propto T^8$ erg/gm/sec. A precise calculation, in the non-relativistic non-degenerate region, gives:

$$\varepsilon = 10^{7\cdot 7}T_9^8 \text{ erg/gm/sec}, \quad T_9 \equiv 10^9\ ^\circ\text{K} \qquad \text{III–2}$$

This rate is hence independent here of the density, but it will become dependent on it in the degenerate region. The effect of the degeneracy is to forbid certain $e^- + \gamma$ collisions (by decreasing the accessible phase space), and hence to decrease the reaction rate.

III-B-2) *The pair annihilation mechanism.* We will now consider the reaction

$$e^+ + e^- \to \gamma + \gamma$$
$$\to \nu + \bar{\nu}$$

(pair annihilation) with the same hypotheses.

The probability of reaction is given here by $P = \sigma v = \pi r_0^2 c = \pi r_0^3 (c/r_0)$ which we can roughly interpret as follows: $(\pi r_0^3/1)$ is the probability of finding the pair at a distance $r_0$ (or less) and $(c/r_0)$ is the inverse of the reaction time. The energy carried by the neutrinos is very close to $2mc^2$ and, for this reason, the ratio $P(\nu\nu')/P(\gamma)$ is more or less constant for all the electron pairs. The energy emitted per unit volume is given by (where $n^+$ is the number of positrons):

$$E = \underbrace{(n^+ n^-)}_{\substack{\text{no. of}\\ \text{pairs}}} \underbrace{\left\{ \sigma v \frac{P(\nu\bar{\nu})}{P(\gamma)} \right\}}_{\substack{\text{prob. of}\\ \text{emission of}\\ \text{a } \nu\bar{\nu}}} \times \underbrace{(2mc^2)}_{\substack{\text{emitted}\\ \text{energy}}} \qquad \text{III-3}$$

In empty space ($\rho = 0$), we would have:

$$n^+ = n^- = (2/\lambda_T^3) \exp\left(-\frac{mc^2}{kT}\right)$$

(calculated by the partition function) $(\lambda_T = h/(2\pi\mu kT)^{1/2})$. This expression is no longer valid in the presence of matter, but (in non-degenerate state) the equation III-3, with this expression, remains valid since the rate of creation $\gamma \to e^+ + e^-$ is not a function of the density and since in equilibrium this rate is equal to the rate $e^+ + e^- \to \gamma$. Exact calculation gives:

$$\varepsilon_{\text{pair}} = \frac{E}{\rho} = 10^{18.7} \exp\left(-\frac{2mc^2}{kT}\right)\left(\frac{T_9^3}{\rho}\right) \quad \text{non-degenerate} \qquad \text{III-4}$$

This process proliferates at low densities and high temperatures. Degeneracy damps it rapidly.

A numerical example illustrates its importance: for $T_9 = 3$, $\rho = 10^6$, $\varepsilon_{\text{pair}} \simeq 10$ eV/nucleon/sec. Now at this time, the available nuclear potential is of several hundred keV (Si to Fe). The nuclear resources will be exhausted in a few days; for $T_9 = 4$, in a few hours!

III-B-3) *Plasma oscillations.* Finally we discuss briefly the emission of neutrinos by plasma oscillations. The interaction of a plasma with an E.M. wave can be analysed in terms of normal modes (plasmons). These normal modes can be de-excited by emission of photons or by emission of neutrinos, with the branching ratio mentioned above.

Because of its collective character, this process is much less affected by degeneracy than the previous ones. It is predominant in high-density and low-temperature regions. In Figs. III–1, III–2, III–3, we can see the intensity of the various processes, and also the iso-intensity curves in the $\rho$–$T$ plane, curves which bound the domain of preponderance of each process.

### III–C: *Neutrinos and the H–R paths*

During the hydrogen burning (M.S.) the only neutrinos emitted come from the nuclear reactions (Lecture II). They represent at most a few per cent of the total energy and influence very little the course of the evolution. Preliminary calculations have shown that the same holds for the red-giant period. There, the predominating mechanism is the plasma oscillation. The emitted neutrinic energy is roughly 10% of the gravitational energy emitted by the contraction of the stellar nucleus.

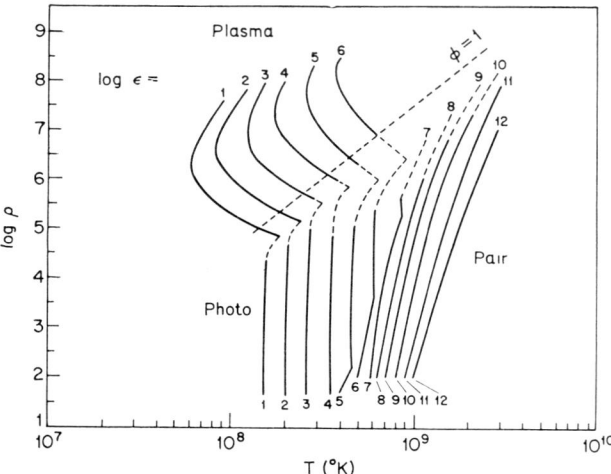

Fig. III–1. Iso-intensity curves of the neutrino emissions in terms of $\rho$ and $T$. For instance, the parameter "1" means that on this curve $\varepsilon = 10$ erg/gm/sec. The words "pair", "photo", "plasma" identify the predominating process in each domain. The neutrinos created by pair annihilation proliferate in regions of high temperature and low density; they are rapidly damped by degeneracy. On the other hand, the plasma neutrinos adapt much better to a reduced degeneracy. The photoneutrinos live in the intermediary domain.

In the neighbourhood of $T_6 \simeq 500$, the neutrino emission is very high. By the virial theorem, in non-degenerate state, the role of the neutrino emission is to raise the temperature of the stages of nuclear burning. Whereas in the absence of neutrinos these stages are approximately characterized by $\varepsilon_N = \varepsilon_{ph}$, now we must have $\varepsilon_N \simeq \varepsilon_\nu \gg \varepsilon_{ph}$ (Lecture II).

Consequently, the fuel burns faster and the path is covered much more rapidly. Consider, for instance, the case of the $h$ and $\chi$ Persei cluster (Fig. III-4). We readily distinguish the group of red giants and the group of red supergiants. It is obvious that the relative population of the two groups must be inversely proportional to the velocity of the points in the corresponding region. The effect of the neutrino emissions is preponderant for the red supergiants. The ratio of the number of elements in each group should allow us to determine the activity (or non-activity) of the neutrinos.

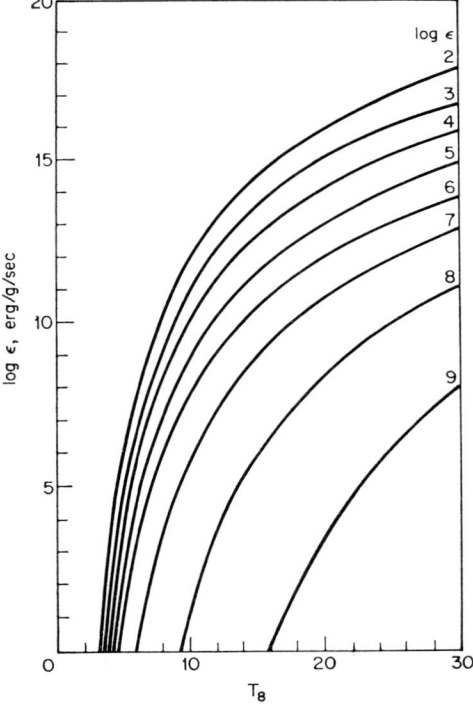

Fig. III-2. Emission of neutrinos by pair annihilation as a function of temperature (abscissa) and of density (curve parameter) (Chiu). We note that at $T_8 = 30$ and $\log \rho = 10^6$ gm/cm$^3$ the emission is greater than $10^{13}$ erg/gm/sec. At such a rate, all the nuclear reserves of a star are dissipated in a few days.

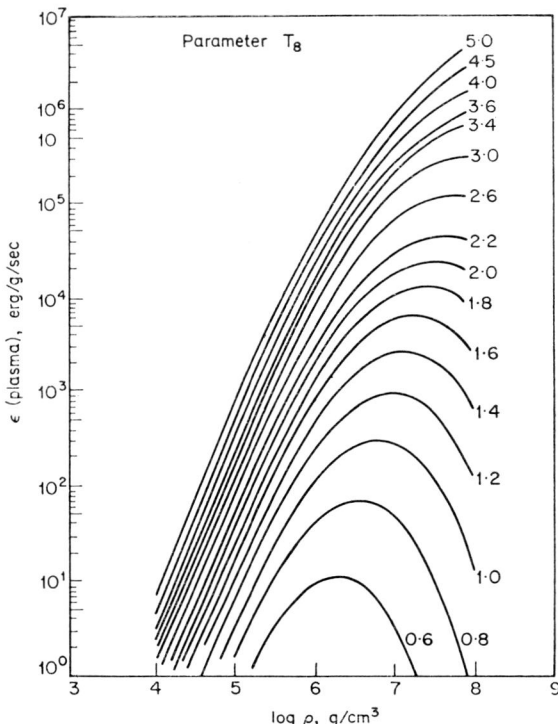

Fig. III–3. Emission of neutrinos by plasma oscillations (Ruderman).

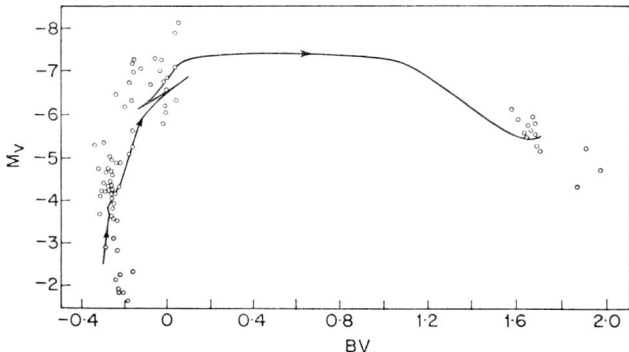

Fig. III–4. Evolutionary path of a $15 \cdot 6 M_\odot$ star inscribed in the H–R diagram of the $h$ and $\chi$ Persei clusters (Hayashi, 1962). To the left, we have the main sequence. Above the point $b$, the heavier stars (roughly $15 M_\odot$) have left the M.S. In the neighbourhood of $c$ they burn the helium and in the neighbourhood of $e$ they burn the carbon.

III–D: *Photodisintegration; equilibrium (e) process; the iron peak*

After the burning of the oxygen, the stellar core is composed mainly of Mg, Si, S. $Si^{28}$ is the most stable of these; the other isotopes photodisintegrate one after the other and finally increase its concentration.

Then when $Si^{28}$ starts disintegrating ($T_6 \simeq 2500$), the exodus towards the region of the most stable elements begins.

The nuclear reactions are extremely rapid so that there settles in, for each element $(A, Z)$, an equilibrium which can be schematized as follows

$$(A, Z) \rightleftharpoons Z \text{ protons} + (A - Z) \text{ neutrons}$$

Always according to statistical mechanics, the abundance ratio is given by the ratio of the partition functions $(K)$

$$\frac{n(A, Z)}{n_n^{A-Z} n_p^Z} = \frac{K(A, Z)}{K_n^{A-Z} K_p^Z} \text{ where } K = \frac{w \, e^{-Q/kT}}{\lambda_T^3}$$

$$w = \sum_j (2J_i + 1) \exp\left(-\frac{E_j}{kT}\right)$$

whence

$$n(A, Z) = \frac{w(A, Z)}{\lambda_{T_A}^3 2^A} n_n^{A-Z} n_p^Z \lambda_{T_p}^{3A} \exp\left\{\frac{Q(A, Z)}{kT}\right\} \quad \text{III-5}$$

where

$$Q(A, Z) = \{M_n(A - Z) + ZMp - M(A - Z)\}c^2 > 0$$

$$\lambda_{T_A} = \frac{h}{(2\pi M_A kT)^{1/2}}$$

is the binding energy of the nucleus $(A, Z)$.

The abundance $n(A, Z)$ depends on three variables: $T$, $n_p$ and $n_n$. We rather use the variables $T$, $n_p/n_n$, and the density $\rho$ which is related to the distribution of the $n(A, Z)$ by

$$\sum M(A, Z) n(A, Z) = \rho \quad \text{III-6}$$

The theory claims to explain the relative abundance of the elements of the iron peak; we normalize thus with respect to $n(56, 26)$

$$\log\left\{\frac{n(A, Z)}{n(26, 56)}\right\} = \frac{\log\{w(A, Z)\}}{w(26, 56)}$$
$$+ \frac{5.04}{T_9}\left\{\frac{Q(A, Z)}{A} - \frac{Q(26, 56)}{56}\right\} + \log\left\{\frac{n_p}{n_n}\left(Z - \frac{26}{56}A\right)\right\}$$
$$+ \frac{A - 56}{56}\left[\log\left\{\frac{n(56, 26)}{w(56, 26)}\right\} - 33.12 - \frac{3}{2}\log T_9 - \frac{3}{2}\log 56\right]$$
$$\text{III-7}$$

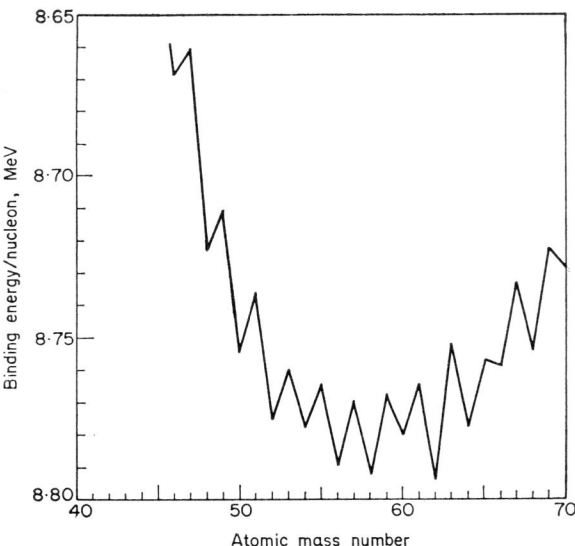

FIG. III–5. Detail of the binding energy curve of the nuclei in the neighbourhood of iron. To an average curve with a definite minimum is added a structure due in particular to the difference of stability between nuclei with even and odd mass number. This influences the lifetimes $t_\beta$ which influence in turn the final distribution of the elements of the iron peak.

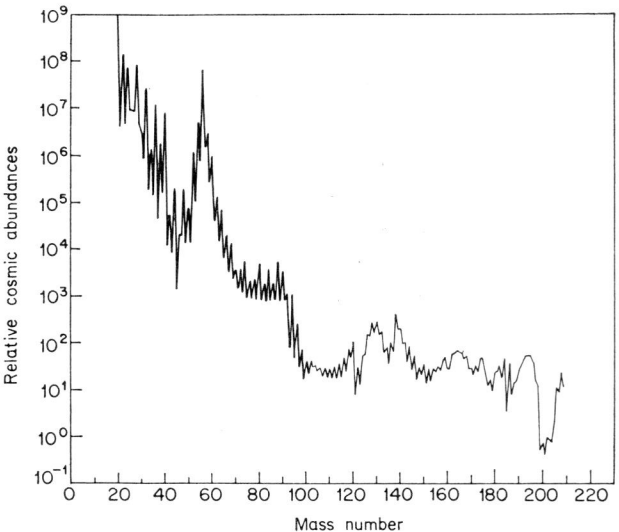

FIG. III–6. Detail of the abundance curve in the neighbourhood of the iron peak. The general appearance of the curve is given by $n_i/n_{\text{Fe}} = e^{-\Delta Q/kT}$, where $\Delta Q$ is the difference in stability between the nucleus $i$ and the iron. The slope is determined by $kT$. The "fine structure" of the curve depends on other factors, such as the pairing energy, the probabilities $t_\beta$ and the duration of the phenomenon.

The most important term is that connected with the temperature. It contains the difference in energy per nucleon between the $(A, Z)$ nucleus and $Fe^{56}$. By means of Figs. I–5 and III–5, we can readily understand the general structure of the iron peak (Fig. III–6).

### III–E: *The time scale of the e process*

The role of the term which contains $n_p/n_n$ is related to the time scale of the phenomenon in the following manner: in the equilibrium process we distinguish between the purely nuclear reactions and the beta reactions. Its importance is due to the fact that the process takes nuclei, which are located in a region of the valley of nuclear stability where the ratio of the number of protons to the number of neutrons $(Z/N)$ is equal to 1 (we then had mostly $Si^{28}$), and carries them to a region where $Z/N \simeq 0.9$ (for instance the stable isotopes of iron). There must hence be one, two or even three $\beta$ disintegrations on the way.

While it is true that we can say that the time scale of the nuclear reactions can be neglected, this is not the case for the times $t_\beta$.

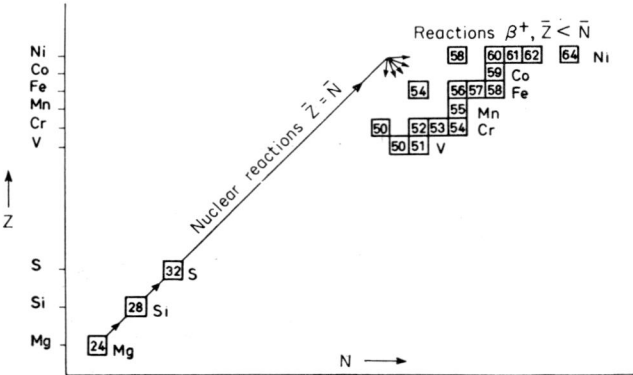

Since the appearance of the iron peak depends on the ratio $n_p/n_n$, we can follow the evolution of the isotopic abundance in terms of the variation of $n_p/n_n$, and determine $n_p/n_n$ when we will have reproduced the observed abundances.

At the beginning of the process, the ratio of the total number of protons to the total number of neutrons is (we mostly have $Si^{28}$, perhaps also a little $Mg^{24}$ and $S^{32}$)

$$\frac{\bar{Z}}{\bar{N}} = \frac{\sum_A Z n(A, Z)}{\sum (A - Z) n(A, Z)} = 1 \qquad \text{III–8}$$

In a very short time, the nuclear reactions give (schematically) $2(_{14}Si^{28}) \rightarrow (_{28}Ni^{56})$. Then, progressively (by $\beta^+$ emission), $\bar{Z}/\bar{N}$ decreases. The value of $n_p/n_n$ at each time $t$ can be calculated by means of equation III–8.

The observed abundances (III–7) (though still very uncertain) indicate $T_6 \simeq 4000$, $\rho \simeq 3 \times 10^6$ and $n_p/n_n \simeq 500$. This last measurement corresponds to $\bar{Z}/\bar{N} \simeq 0.87$. This is not the value we would

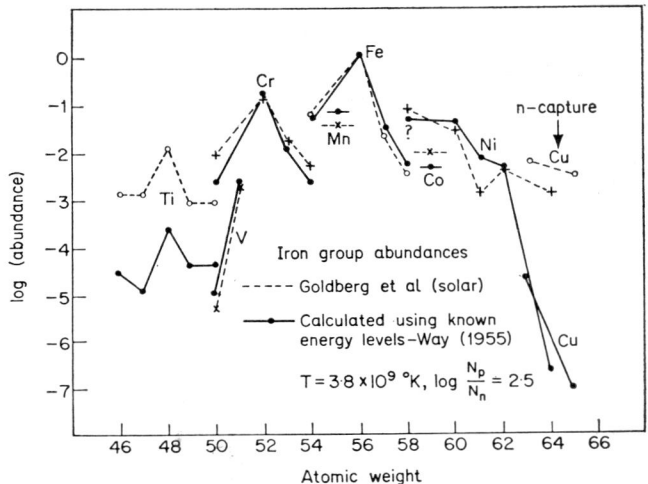

Fig. III–7. Theoretical calculation of the abundances of the elements of the iron peak. Choosing $T_6 = 3800$ and $n_p/n_n = 300$, we get good agreement from $A = 50$ to $62$. The disagreement above $A = 62$ can be explained by the later absorption of neutrons (Lecture IV). The disagreement below 50 has still not been explained ($B^2FH$).

obtain if the phenomenon had lasted long enough to allow equilibrium between the $\beta^+$ and $\beta^-$ reactions. Now, such an equilibrium settles in in a few days (the calculations are based on the systematics of the nuclei in question).

The difference between the observed $\bar{Z}/\bar{N}$ and $\bar{Z}/\bar{N}$ at the ($\beta^+ = \beta^-$) equilibrium allows us to calculate the length of the $e$ process; some ten hours at most.

### III–F: *Neutrinos and the iron peak*

In the preceding lesson, we studied the carbon- and oxygen-burning stages by means of energy models defined by $\varepsilon_\nu = \varepsilon_N$. The same models can be used here. In going from $Si^{28}$ to $Fe^{56}$ the stellar core releases

$\simeq 1$ MeV/nucleon, hence $\simeq 10^{18}$ erg/gm. At $T_6 = 4000$, this energy is emitted in a few hours by the neutrinos (pair annihilation). On the other hand, in the absence of neutrinic processes, the dissipation of energy by photon channels would take a few thousand years, under these conditions. The $\beta^+ = \beta^-$ equilibrium would be reached and the abundances would be different.

The distribution of elements in the iron peak seems to indicate the existence of neutrinic processes.

III-G: *The end of the sequences of states of gravitational equilibrium*

If the temperature continues to rise, the activity of the neutrinos and of the gammas will lead us to a doubly implosive situation; at $T_6 \simeq 5000$, the neutron emission is so intense that the gravitational contraction must accelerate prodigiously in order to keep up with the energy release. When the velocity of contraction reaches the so-called

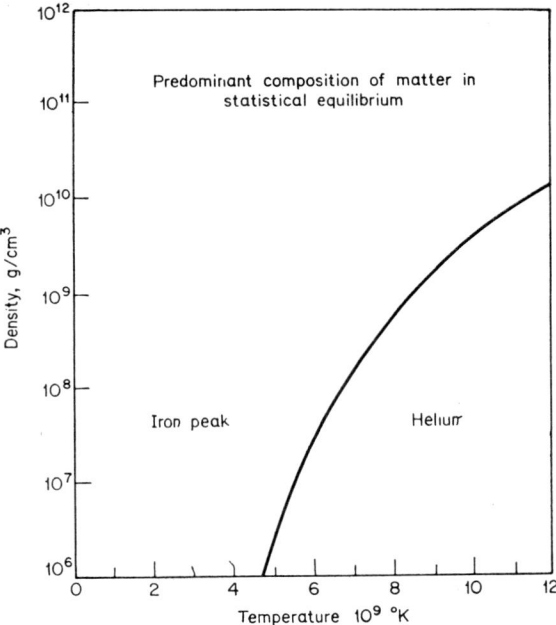

FIG. III-8. In the left-hand side region of the diagram, the statistical equilibrium favours the iron peak. In an extremely narrow region around the curve, we go suddenly from Fe to He by a series of successive photodisintegrations. The phenomenon is violently endothermic and very harmful to the hydrodynamic equilibrium of a star (Cameron, 1963).

free-fall velocity, we have implosion. In the same range of temperature, the $Fe^{56}$ which is concentrated at the centre of the star becomes unstable (Fig. III-8). It disintegrates rapidly and endothermically ($-2\cdot 2$ MeV/nucleon) into $13He^4$ and $4n$. The stellar nucleus, deprived of its heat, would then collapse.

Whatever its actual cause may be, the effect of the implosion is to project into the stellar nucleus the layers of fuel from the stellar envelope (Si, O, C, He, H). These fuels suddenly take fire at the temperatures of the stellar nucleus. We would get a huge explosion.

### III-H: *Supernovae (S.N.)*

With a frequency of about 1 per 300 years, we observe in the galaxies the appearance of an object which reaches, for a few hours, $10^9$ solar luminosities or $\simeq 10^{43}$ erg/gm/sec. We distinguish two types of supernovae: Type I is composed of small ($\sim 1M_\odot$) and old (Pop. II) stars; after the maximum its light curve has an exponential form, with a lifetime varying from 25 to 70 days, the ejected mass is $\simeq 0\cdot 1M_\odot$. Type II

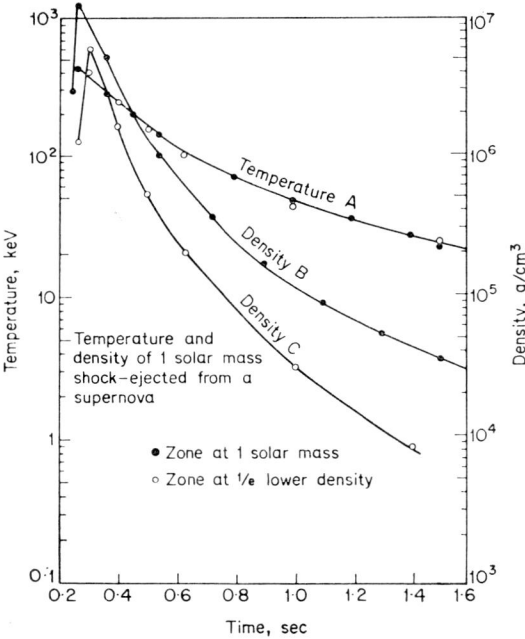

FIG. III-9. Profile of the density and the temperature (1 keV $\simeq 10^7$ °K) as functions of time, in the surface regions ejected from a supernova (Colgate, 1962).

is composed of large ($\sim 20 M_\odot$) and young (Pop. I) stars, the light curve has no particular form, the ejected mass is equal to several solar masses.

The evolution of the future S.N. of Type II is probably not influenced by the degeneracy effect. Consequently, the temperature must reach the implosive domain ($T_6 \simeq 5000$). One is tempted to explain the S.N. II by the sequence of events described in the preceding section.

For S.N. I, on the contrary, degeneracy must play an important role. A phenomenon characteristic of degenerate systems would be the nuclear "flash" (Lecture II). Are the S.N. I caused by giant "flashes"?

Unlike the physical conditions which immediately precede the S.N. explosion, which have hardly been studied until now, the propagation of the shock wave at the time of the explosion has been the object of many studies. Fig. III-9 illustrates the propagation of the shock wave at two different places of the stellar mass.

The phenomenon of restitution of stellar matter is well illustrated in Figs. I–6 and I–7. Part of the star probably falls back onto itself, and becomes a white dwarf, or perhaps a neutron star. Neutron stars are perhaps visible by means of an X-ray telescope.

## Bibliography

A good start:

L. D. Landau, E. M. Lifshitz, "Statistical Physics", Pergamon Press (1958)
E. J. Konopinski, *Ann. Rev. of Nucl. Sciences*, **9**, 99, 1959
R. P. Feynmann, M. Gell-Mann, *Phys. Rev.* **109**, 193, 1958

Direct coupling of electrons and neutrinos, suggested by:

B. M. Pontecorvo, *J.E.T.P.* **9**, 1148, 1959

is studied, among others, in:

H. Y. Chiu, P. Morrison, *Phys. Rev. Lett.* **5**, 573, 1960
H. Y. Chiu, *Phys. Rev.* **123**, 1040, 1961
J. B. Adams, M. A. Ruderman, C. H. Woo, *Phys. Rev.* **129**, 1383, 1963

Astrophysical developments and speculations:

W. A. Fowler, J. L. Vogl, "Nuclear and Neutrino Processes in Stars and Supernovae", Lectures in Theoretical Physics, Vol. VI, University of Colorado Press (1963)
W. A. Fowler, F. Hoyle, *Ap. J.* Suppl. 1965
P. Ledoux, *Astrophysica Norvegica*, Vol. IX, 187, 1964
H. Reeves, *Ap. J.* **138**, 19, 1963
H. Y. Chiu, *Ann. Phys.* **15**, 1, 1961

The shock wave in a supernova:

S. A. Colgate, A. G. W. Cameron, *Nature*, **200**, 870, 1963
S. A. Colgate, M. H. Johnson, *Phys. Rev. Lett.* **5**, 235, 1960

# IV

## The role of neutrons; heavy elements synthesis

### IV-A: *Why neutrons?*

THE earlier theories of nucleosynthesis attributed to the phenomenon of the absorption of neutrons by lighter elements the existence of all the elements heavier than hydrogen. Neutrons appeared at the beginning of the universe, in appropriate amounts. The absence of stable elements of mass 5 and 8 soon turned out to be an unsurmountable obstacle. As we have seen earlier, the production of the lighter elements ($1 < A \leqslant 40$) is now attributed to the capture reactions between charged particles, and to the reactions of photodisintegration and of recombination in statistical equilibrium the production of the elements contained in the iron peak, $40 < A < 65$.

The heavy elements ($A > 65$), because of their high charge and the relative weakness of their stability, can be explained by neither of these two processes. It is hence natural to reintroduce the hypothesis of neutron absorption by the light elements which are already formed, in particular by the iron-peak elements.

This hypothesis is corroborated in several ways: on the one hand, a detailed study of the chain of nuclear reactions which take place during stellar evolution shows that at certain times, large neutron fluxes are released in the core of the star. On the other hand, the analysis of the relative abundance of the elements formed in this way shows certain regularities which can only be explained in terms of neutron absorption. We will consider these two topics separately.

### IV-B: *Irradiation parameters*

Before describing the phenomena which give rise to the release of neutrons, it is useful to obtain a method for calibrating these sources, in order to be able to evaluate their relative nucleosynthetic importance.

First, we obtain the number of neutrons produced with respect to a given mother-isotope. From this we deduce the number of neutrons per iron nucleus (in all likelihood the seed-nucleus of the heavier atoms). The fraction ($f$) of the neutrons absorbed by each of these nuclei is given in reality by

$$f = \frac{\langle \sigma v \rangle_{56} n_{56}}{\sum_A \langle \sigma v \rangle_A n_A} \qquad \text{IV-1}$$

In the absence of notorious poisons (e.g. $He^3$, $N^{14}$), this fraction is close to unity. To make things easier, we set $f = 1$.

It is also interesting to obtain the density of neutrons in thermal equilibrium during a stage of nuclear evolution, the flux of neutrons per surface unit, as well as the integral of the neutron flux with respect to time, the order of magnitude of the average time between the emission of each neutron, and finally the lifetime of a given heavy isotope with respect to the neutron absorption. These last quantities depend on the density and the temperature of the stellar medium. These parameters must be obtained from an already-known stellar model. Since, on the other hand, the neutronic phenomena influence the evolution only to a negligible extent, we will not have to worry about their effect on the evolution (feed-back). Figs. II–20 and II–21 give examples of the evolutive sequences of the temperature and the central density for the phases of hydrogen, helium and carbon burning.

In statistical equilibrium, the number of neutrons per volume unit, $n_n$, is given by

$$\frac{dn_n}{dt} = n_m n_c \langle \sigma v \rangle_{m,c \to n} - \sum n_A \langle \sigma v \rangle_{A, n \to \gamma} n_n = 0$$

$$\therefore n_n = \frac{n_m n_c \langle \sigma v \rangle_{m,c \to n}}{\sum n_A \langle \sigma v \rangle_{A, n \to \gamma}} \qquad \text{IV-2}$$

We considered here a neutron-producing reaction

$$\text{m} + \text{c} \to \text{n} + \ldots$$
$$\searrow \ldots$$

and absorption reactions

$$\text{n} + \text{A} \to \gamma + \text{A} + 1$$

The $\langle \sigma v \rangle$ were defined in Lecture II, equation II–4. The term

$\langle \sigma v \rangle_{A,n \to \gamma}$ describes the capture of neutrons by $(n, \gamma)$ reactions on heavy elements. The cross-section $\sigma(n, \gamma)$ can be written in the form:

$$\sigma(n, \gamma) = 2\pi^2 \lambda^2 \sum_j \frac{w_j \Gamma_n^j \Gamma_\gamma^j}{(\Gamma_n^j + \Gamma_\gamma^j) D^j} \qquad \text{IV-3}$$

The stellar temperatures correspond to average energies $kT$ lower than a few hundred keV. For such energies, the $s$ ($l = 0$) and $p$ ($l = 1$) incident waves are preponderant. The cross-sections vary rather slowly with energy $\sigma \propto E^{-1/2}$ or $E^{-1}$, and the value of $\langle \sigma v \rangle(T)$ is fairly well given by $\sigma(T) v(T)$ or $\sigma(T) = \sigma(E = kT)$ and $v(T) = (2kT/m)^{1/2}$. For a given element $A$, the product $\sigma v$ decreases in general rather slowly with temperature. This decrease is more or less described by

$$\sigma v = \sigma v(T_0)(T_0/T)^{0.8 \text{ or } 0.9}$$

The neutron flux is given by $n_n v(T)$.

The variable $\Delta \tau$ is defined (in $10^{29}$ units) by: $\int n_n v(T) \, dt$. We associate a $\Delta \tau$ to each neutron-producing reaction. The average time between each neutron emission is given by $t_{m, c \to n} = [n_m n_c \langle \sigma v \rangle_{m, c \to n}]^{-1}$, and the lifetime of a heavy isotope $(A)$ with respect to neutron absorption is $t_{A,n} = [n_n \langle \sigma v \rangle_{A,n}]^{-1}$. In the cases which interest us, we find $v(T) \simeq 3 \times 10^8$ cm/sec, and $\sigma \simeq 100$ to $1000$ mb ($1$ mb $= 10^{-27}$ cm$^2$). Hence $t_{A,n} \simeq 10^{16}/n_n$ (sec). *Note the definition of $t_{A,n}$; it is not the average life of the neutron* (12 minutes)!

Finally, we consider the integrals $\langle \sigma v \rangle_{m, c \to n}$ of neutronic branching channels. We obtain them by calculating the capture rate $\langle \sigma v \rangle_{m,c}$ as we did earlier (Lecture II), and multiplying it by the branching ratio

$$\langle \sigma v \rangle_{m,c \to n} = \langle \sigma v \rangle_{m,c} g_n$$

$$g_n = \frac{\int_{E_t}^{\infty} \sigma_{m,c}(E) n(E) v \, dE (\Gamma_n/\Gamma)}{\int_0^{\infty} \sigma_{m,c}(E) n(E) v \, dE} \qquad \text{IV-4}$$

$\Gamma_n/\Gamma$ is here the ratio of the neutron width to the total width of the corresponding excited levels, and $E_t$ is the threshold energy of the neutron emission. Since the ratio $\Gamma_n/\Gamma$ varies only in a very small section of the integration region (for very low energies of the neutron), we can consider it as a constant and remove it from the integral. We write $g = \overline{(\Gamma_n/\Gamma)} F$.

In first approximation, $F = 0$ if $E_t > E_0$; $F = 1$ if $E_t < E_0$. The "threshold" of neutron emission is reached at temperature

$$kT_n = \frac{\sqrt{2}\hbar E_t^{3/2}}{\pi e^2 Z_1 Z_2 \sqrt{M_p}}$$

By approximate integration, we obtain

$$F = (\tfrac{1}{2}x_t^{5/4})\{1 - \phi(z_t)\} \exp\{\tfrac{1}{9}\tau(x_t^{5/2} - 5x_t - 5x_t^{-1/2} + 9)\} \qquad \text{IV-5}$$

for $x_t > 1$. Here

$$x_t = \frac{E_t}{E_0}, \quad z_t = \tfrac{1}{3}\sqrt{\tau}(x_t^{5/4} - x_t^{-1/4})$$

$$\phi(z_t) = \int_0^{z_t} \left(\frac{2}{\sqrt{\pi}}\right) \exp(-z^2) dz$$

The other terms, such as $E_0$, the Gamow energy, and $\tau$, were defined in Lecture II (this $\tau$ is not the irradiation!).

For the cases we are interested in ($x_t \geqslant 1$), the following approximation is acceptable: $F = \{1 - \phi(z_t)\}/2$.

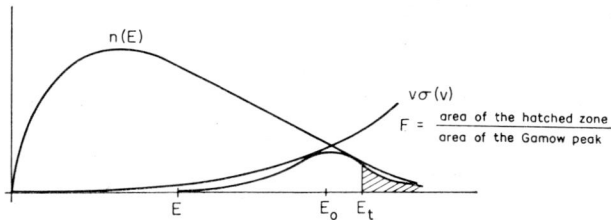

### IV-C: *Neutron sources in stars*

IV-C-1) *Before the hydrogen-burning phase.* The deuterium present in the contracting stellar gas burns in the neighbourhood of $T_6 = 1$. The relative abundance of deuterium at that time is very little known; it is smaller than $10^{-4}$ (compared with the hydrogen) but might be very much smaller. For the time being, we choose the larger value.

The deuterium is destroyed by the D(p, $\gamma$)He$^3$, D(D, n)He$^3$ and D(D, p)H$^3$ reactions. From these reactions we obtain 0·1 neutron per initial deuteron, hence $10^{-5}n$ per hydrogen atom, and $\simeq 1$ neutron per iron-peak atom (Pop. I : $n_{56}/n_H \simeq 10^{-5}$). Moreover, the presence of He$^3$ reduces still further the number of neutrons available for the synthesis of heavy elements (the sequence He$^3$(n, p)H$^3 \rightarrow$ He$^3$ shows that He$^3$ is not eliminated, but serves as a catalyser for the reconversion of the neutrons into protons). For practical purposes, it seems we can neglect the contribution of the deuterium towards the synthesis of the heavy elements.

The hydrogen-burning phase does not produce any neutrons.

IV–C–2) *The contraction phase preceding the helium burning* $C^{13}(\alpha, n)O^{16}$. The nuclei of $C^{13}$ are produced during the phase of hydrogen burning. This phase transforms the whole CNO group into $\simeq 0.04 C^{12}$, $\simeq 0.01 C^{13}$ and $\simeq 0.95 N^{14}$. Even for a star of Population I, $n_{13}/n_{\text{nucleon}} \simeq 10^{-5}$, hence $n_{13}/n_{\text{Fe}} \simeq 1$. Moreover, by examining Fig. IV–1, we see that $C^{13}$ always burns before $N^{14}$ and hence in the presence of $N^{14}$. By the $N^{14}(n, p)C^{14} \to N^{14}$ reaction, nitrogen is a neutron poison. However, if the p produced in this way are captured by $C^{12}$, we have

$$C^{12}(p, \gamma)N^{13} \to C^{13}(\alpha, n)O^{16}$$

and the neutron is regenerated. In other words, $C^{12}$ can actually be an antidote to $N^{14}$. Will this happen in this case? No, because at the end of the carbon cycle the statistical equilibrium between captures could be expressed by the following expression, which is in fact responsible for the abundances mentioned earlier:

$$n_{12}\langle\sigma v\rangle_{12,1} = n_{13}\langle\sigma v\rangle_{13,1} = n_{14}\langle\sigma v\rangle_{14,1} = n_{15}\langle\sigma v\rangle_{15,1} \qquad \text{IV–6}$$

so that a small fraction of the protons actually caught on to $C^{12}$. Notwithstanding the increase in temperature since the end of the stage of helium burning, the equation remains relatively valid, and $C^{12}$ cannot fulfil its antidote functions. Therefore we will also neglect this mechanism.

We must mention here an analogous mechanism which has been

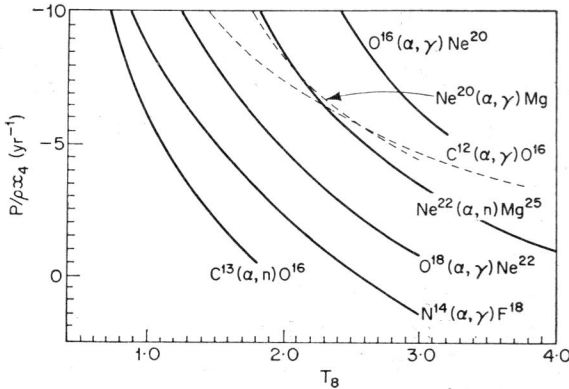

Fig. IV–1. Rate of thermonuclear reactions of various light isotopes with helium ($T^8 = 10^8$ °K), calculated by the methods described in Lecture II.

recently considered to explain certain abundance anomalies at the surface of CH stars. We assign to the helium "flash" (Lecture II) the task of mixing to the helium core of a star a certain amount of hydrogen. Then we consider the sequence $C^{12}(p, \gamma)N^{13} \to C^{13}(\alpha, n)O^{16}$. We note that this reaction occurs at the beginning of the helium burning, hence after the vanishing of the $N^{14}$ by $N^{14}(\alpha, \gamma)O^{18} \to F^{18}$. The proton absorption will trigger a new CNO cycle which could eventually cause the reappearance of $N^{14}$, but the massive injection of $C^{12}$ allows this isotope to fulfil its antidote function.

The flash occurs at the very beginning of the helium burning. The $C^{12}(\alpha, \gamma)O^{16}$ reaction has not yet started.

The main difficulty comes from the hypothesis of the mixing of the envelope with the core of the star. The existing stellar models seem rather forbidding. The possibility exists, but it is marginal.

On the other hand, the agreement between the theory and the observations is quite remarkable; we observe indeed on the CH stars an increase in C and the *heavy elements* without a simultaneous increase in O. The hypothesis should not be discarded.

IV-C-3) *Helium burning.* The nitrogen 14 present in the stellar core at the end of the CNO cycle is transformed during the helium burning in the following manner:

$$N^{14}(\alpha, \gamma)F^{18}(e, \nu)O^{18} \to O^{18}(\alpha, n)Ne^{21} \quad Q = -700 \text{ keV}$$
$$\searrow O^{18}(\alpha, \gamma)Ne^{22}(\alpha, n)Mg^{25} \quad Q = -482 \text{ keV}$$
$$Ne^{22}(\alpha, \gamma)Mg^{26}$$

perhaps followed by

$$Mg^{25}(\alpha, n)Si^{28}$$
$$Mg^{26}(\alpha, n)Si^{29}(\alpha, n)S^{32}$$

Hence we have here two endothermic reactions which *can* produce neutrons. In the first case, the $O^{18} + He^4$ capture gives $(n + Ne^{21})$ if $T > T_n = 350$ ($\overline{\Gamma_n/\Gamma} \simeq 1$). The problem is to prevent the $O^{18}$ from burning before this temperature is reached. The reheating must be very sudden, as in the helium flash. The models we have at present do not quite reach high enough temperatures however (250 to 300 at most).

The second reaction, $Ne^{22}(\alpha, n)Mg^{25}$ ($\overline{\Gamma_n/\Gamma} = 1$) has a $T_n = 200$. Will the sequence of captures indicated in the diagram reach $Ne^{22}$ before exhaustion of the $\alpha$, and even, if it does, will the $Ne^{22}$ burn before we reach $T_n = 200$ (hence without emitting neutrons)? We first see that during the helium burning the $C^{12}(\alpha, \gamma)O^{16}$ becomes fairly rapidly the main consumer of the α. From the diagram IV–1, it is clear that the reactions $N^{14}(\alpha, \gamma)F^{18}$ and $O^{18}(\alpha, \gamma)Ne^{22}$ are always more rapid than $C^{12}(\alpha, \gamma)O^{16}$; $Ne^{22}$ will be formed before exhaustion of the helium. Again according to the diagram, as long as $T_6 < 200$, $C^{12}$ absorbs alphas more rapidly than $Ne^{22}$. Hence we have the following scheme: From 100 to 200, $Ne^{22}$ is formed and partially preserved; at $T_6 > 200$, $Ne^{22}$ burns, emitting a neutron. From the models we know, a star of several solar masses will reach (and exceed) the temperature $T_c = 200$ before the helium is exhausted (cf. Fig. I–21). The neutron concentration will reach $\simeq 10^5 n/cm^3$, and the lifetime of an iron atom with respect to the neutron absorption will be $\simeq 10^{11}$ seconds. The number of neutrons emitted per iron nucleus could be as high as ten and the parameter $\Delta\tau = 0{\cdot}1$ (in units of $10^{27}$).

IV–C–4) *Carbon burning.* $C^{12} + C^{12} \rightarrow Mg^{23} + n$, $Q = -2{\cdot}603$ MeV and $\Gamma_n/\Gamma \simeq 0{\cdot}05$. We meet for the first time a source of neutrons obtained directly from the energy-producing reaction. The branching is endothermic ($T_n \simeq 1200$). In the table, we have, for several temperatures, the number of neutrons emitted per initial $C^{12}$ atom, then the number of neutrons per iron atom. (We have taken $X_{12} = 0{\cdot}5$, $X_{56} = 10^{-3}$ (Pop. I) or $10^{-4}$ (Pop. II).) We neglected the production, by the reaction $C^{12} + C^{12}$, of isotopes such as Mg which, towards the end of the burning, might absorb quite a few of the neutrons.

TABLE

| $T_6$ | 900 | 1000 | 1100 | 1200 |
|---|---|---|---|---|
| $N_n/N_{12}$ | $3 \times 10^{-3}$ | $5 \times 10^{-3}$ | $10 \times 10^{-3}$ | $12 \times 10^{-3}$ |
| $N_n/N_{Fe}$ (Pop. I) | 5 | 10 | 20 | 25 |
| $N_n/N_{Fe}$ (Pop. II) | 50 | 100 | 200 | 250 |

The number of neutrons is a function of the effective burning temperature, hence of the stellar mass. As we have seen in Lecture III, the hypothesis of neutrinic dissipation of the stellar energy seems to assign sufficient temperatures to the carbon burning to make it an important source of neutrons; at $T_6 = 1000$, $n_n = 10^{11}$ cm$^3$, the capture of neutrons by heavy elements takes $\simeq 10^5$ seconds and $\Delta\tau = 0{\cdot}1$ ($10^{27}$ units).

IV–C–5) *Oxygen burning.* $O^{16} + O^{16} \rightarrow Si^{31} + n$, $Q = 1{\cdot}459$ MeV. Here the reaction is exothermic. The branching ratio (very little known) is probably close to 1/10.

The number of neutrons per $O^{16}$ atom is roughly $0{\cdot}05$, so that here we could easily obtain about a hundred neutrons per iron atom. At $T_6 = 1600$, $n_n = 10^{12}$, $t_{n,\gamma} \simeq 10^4$ sec and $\Delta\tau \simeq 1$.

It is still not known whether the burning phases of the carbon and of the oxygen give rise, in small mass stars, to stellar flashes. If they do, the neutron densities could reach $n_n = 10^{18}$ to $10^{20}$ and $t_{n,\gamma} < 1$ sec. We will retain this possibility.

IV-C-6) *Later sources.* Do the later periods produce neutrons? The equilibrium processes produce a large number of them (Lecture III), but on the other hand they mercilessly destroy all the heavy elements. The S.N. explosion probably produces neutrons, at the centre by the reaction $\gamma + Fe^{56} \to 13He^4 + 4n$ and in the envelope, for instance, by $(\gamma, n)$ generated by the shock wave. The first source is still located at too high a temperature, the heavy elements are still destroyed. We cannot say very much about the second source, except that it will not last long (seconds), and will be at temperatures higher than a billion degrees. The densities $n_n$ would be very high and the times $t_{n,\gamma}$ very short.

IV-C-7) *Source characteristics.* Summing up, we retain the following sources: $C^{12} + H$ (convection) at $T_6 \simeq 100$, $N^{14} \to Ne^{22}(\alpha, n)$: $T_6 > 200$, $N^{14} \to O^{18}(\alpha, n)$: $T_6 > 350$, $2C^{12} \to Mg^{23} + n$: $T_6$ from 900 to 1100, $2O^{16} \to Si^{31} + n$: $T_6$ from 1400 to 2000, supernovae: $T_6$ from 1000 to $\simeq 3000$. We note that these sources come into action only at relatively well-determined temperatures. Conversely, if we succeed in determining by some other method the ambient temperature at the time of synthesis of the heavy nuclei (for instance, by studying the observed distributions), we could identify at least partially the nature of the generating reaction. In fact, we could rather think that a given seed-nucleus has been subjected to several irradiations $\Delta\tau_i$, having successively lived in the interior of different stars, and having perhaps received in each one of them the successive irradiations of the different reactions ($\Delta\tau_{Ne^{22}}$, $\Delta\tau_{C^{12}+C^{12}}$, etc.). We would hence have an irradiation distribution giving rise to the observed elements.

The absorption times of neutrons by the heavy elements $(t_{n,\gamma})$ are relatively long (days or years) for all the reactions associated with periods of quiet evolution. Much shorter times can occur only during the flashes or during the final cataclysms.

### IV-D: *Heavy elements*

The abundance of elements with $A > 60$ is illustrated in Fig. IV-2. All the elements beyond Bi ($Z = 83$) are unstable. However, in the region $Z = 92-94$, the periods of disintegration remain sufficiently long

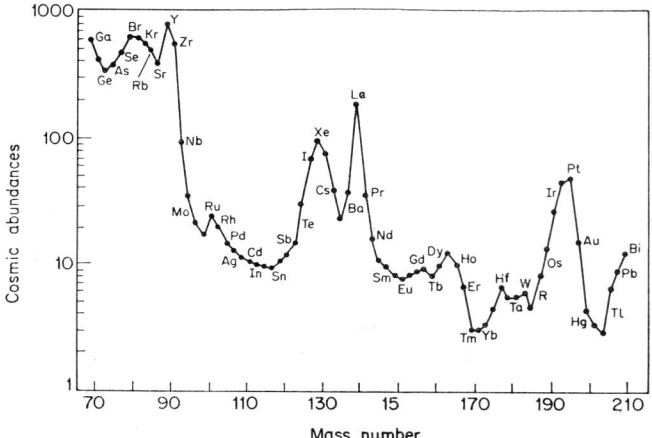

Fig. IV-2. Abundance of heavy elements.

for one to be able to find again in nature, for instance, the isotopes $U^{235}$, $U^{238}$, $Th^{232}$. In Fig. IV-2, we distinguish, on a generally decreasing envelope, the presence of three pairs of peaks, coupled two by two (Br, Y), (Xe, La), (Pt, Bi). Other characteristics of this isotopic distribution appear when we divide these elements into three classes, a division suggested by Fig. IV-3.

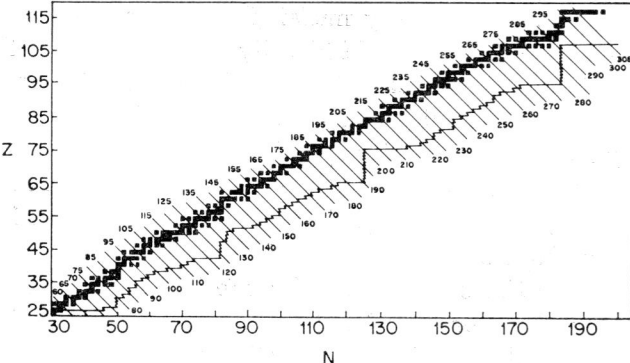

Fig. IV-3. The $Z$–$N$ plane of the heavy elements. The black squares are the stable elements; we have a valley of nuclear stability. At the bottom of the valley (nearly continuous line), the $s$ elements. To the right and to the left (isolated squares) the $p$ and $r$ elements. Further to the right, the thin line represents the path of the $r$ process (Clayton, 1962).

The position of the stable elements in the $Z$–$N$ plane is represented by a black square. We see that quite a few isobars ($A$ = constant) have up to three stable members. This phenomenon can be explained in terms of nuclear physics by the competition between the Coulomb energy and the coupling energy in establishing the nuclear mass. If we raise a vertical axis above the $Z$–$N$ plane, along which we measure the nuclei masses, we obtain a three-dimensional structure characterized by the presence of a more or less diagonal valley. The diagram illustrates the situation by a vertical cross-section along a line $A = Z + N =$ even. The elements located at the bottom of the valley are called the $s$ elements. They form a more or less continuous sequence in Fig. IV–3.

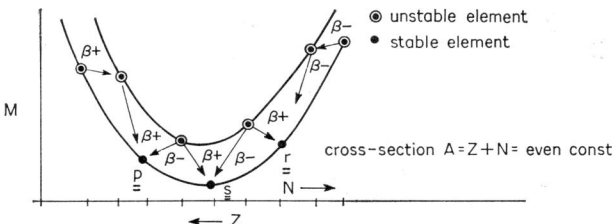

The isobars located to the right of the valley (the so-called $r$ elements) are comparatively richer in neutrons than the $s$ elements. Their total abundance compares more or less with that of the $s$ elements. Finally, the inhabitants of the other side (to the left of the stability valley) are called $p$ (rich in protons). They are in general from 100 to 1000 times rarer than the $s$ or $r$ elements (Fig. I–9). The position of these groups in the $Z$–$N$ plane requires different formation conditions (Fig. IV–4). Consider the path obtained when, for instance, the $_{48}Cd^{110}$ is subjected to a neutron flux. By $(n, \gamma)$ absorption, the path leads us to $_{48}Cd^{115}$ which is unstable: it beta-disintegrates in 54 hours into $_{49}In^{115}$. If the

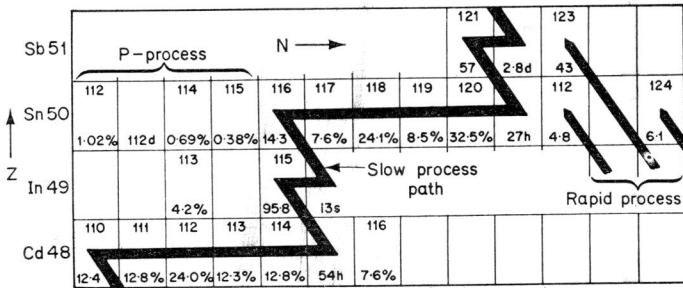

Fig. IV–4. The path of the $s$ process. The horizontal displacements are caused by the $(n, \gamma)$ reactions and the oblique displacements are caused by the $\beta^-$ disintegrations. We note that the $p$ elements (such as $Sn^{112}$) or the $r$ elements (such as $Sn^{124}$) are never produced by the $s$ process (Clayton, 1962).

lifetime of $_{49}Cd^{115}$ with respect to the neutron absorption exceeds this value, we will mostly obtain $_{49}In^{115}$, otherwise we will mostly have $_{48}Cd^{116}$. Here the critical criterion is the ratio $t_{n,\gamma}/t_\beta$. (Most of the times $t_\beta$ are between 0·1 s. and several years.) If $t_{n,\gamma}/t_\beta \gg 1$, the path will constantly follow the bottom of the valley of nuclear stability; the $s$ (slow) elements will result from it, and the $r$ elements will always be bypassed. If $t_{n,\gamma}/t_\beta \ll 1$, the path will turn very much to the right, then, when irradiation stops, the nuclei will emigrate towards the valley and will form, among others, the $r$ (rapid) elements. We note that the $s$ path must stop at $A = 209$ (Pb). Above this limit, the elements are unstable with respect to $\alpha$ disintegration: in the region from $A = 210$ to 220, the lifetimes are very short.

## IV–E: *The s elements*

A correlation between the abundance of the $s$ elements and the cross-section for the $(n, \gamma)$ neutron capture of these elements has been established many years ago. In particular, one noted very early the large abundance in nature of isotopes containing magic numbers of neutrons (isotopes which have particularly small $(n, \gamma)$ sections): $N = 50$ ($Sr^{88}$, $Y^{89}$, $Zr^{90}$); $N = 82$ ($Ba^{138}$, $La^{139}$, $Ce^{140}$, $Pr^{141}$); $N = 126$ ($Pb^{126}$, $Bi^{126}$). These are the right-hand side members of the pairs of peaks in Fig. IV–2. In fact, if we multiply the product of the abundance of the $s$ elements

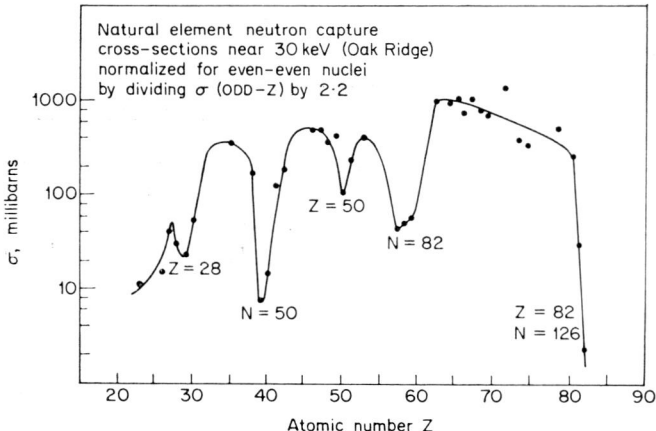

Fig. IV–5. Cross-section of the heavy elements in the neighbourhood of 30 keV for $(n. \gamma)$ capture reactions. We note the very pronounced minima in the neighbourhood of the magic numbers of neutrons. The presence of these minima partially motivated the epithet "magic" given to these nuclei, at the beginnings of the study of nuclear physics (Oak Ridge data).

($N_s$) with the $(n, \gamma)$* cross-section of each of these elements (Fig. IV-5), we obtain a curve which is remarkably continuous (Fig. IV-6).

A similar curve, obtained with the $r$ elements (Fig. IV-7), shows a completely different result. The regularity of the $\sigma_{n,\gamma} N_s$ curve is probably significant. What is the meaning of this correlation.

We have assumed that the formation of the $s$ elements requires the inequality $t_{n,\gamma}/t_\beta \gg 1$, hence only the $t_{n,\gamma}$ are important. The variation in the abundance of an element $A$ is given by: $t_{A,n \to \gamma} = (n_n \langle \sigma v \rangle_{A,n \to \gamma})^{-1}$.

$$\frac{dN_A}{dt} = \langle \sigma v \rangle_{A-1, n \to \gamma} n_n N_{A-1} - \langle \sigma v \rangle_{A, n \to \gamma} n_n N_A \qquad \text{IV-7}$$

If the irradiation had lasted long enough for equilibrium to be reached, we would have:

$$\frac{dN_A}{dt} = 0 \quad \therefore \quad N_{A-1} \langle \sigma v \rangle_{A-1, n \to \gamma} = \langle \sigma v \rangle_{A, n \to \gamma} N_A = \text{constant} \qquad \text{IV-8}$$

The monotonicity of the curve in Fig. IV-6 shows that equilibrium was in the process of settling in; in small intervals $\Delta A$ we have $dN_A/dt \simeq 0$.

This correlation between the observed $s$ abundances and the cross-sections measured in the laboratory is one of the strongest proofs of neutronic activity in the synthesis of the heavy elements. The absence of fluctuation in Fig. IV-6 compared with the rapid variations (a factor of ten or one hundred) in Figs. IV-2 and IV-5, is altogether remarkable.

More information can be deduced from the curve IV-6. Theoretically we should be able to determine the temperature, or the temperatures, at the time of the neutron absorption. Since, at least for some isotopes, $\langle \sigma v \rangle (T)$ varies fairly well with $T$, we could try to find which $T$ gives the most continuous curve IV-6. The cross-sections in the corresponding domain ($\sim 5$ to 100 keV) have not yet been determined with enough precision. Several groups are working at it.

The curve in Fig. IV-6 can also give us information on the mode of irradiation of the seed-nuclei. We first consider the isotopic (fractional) distribution obtained by the irradiation of a unique Fe nucleus in terms of the intensity $\Delta \tau$ of the irradiation, which amounts to solving for all $A$ the equation

$$\frac{dN_A}{d\tau} = (\psi_{A-1} - \psi_A) \qquad \begin{aligned} d\tau &= n_n v \, dt \\ \psi_A &= (\sigma_A N_A) \end{aligned} \qquad \text{IV-9}$$

with the initial conditions $N_{56} = 1$, $N_A = 0$ for $A > 56$.

* The cross-sections are measured at an energy of 30 keV, hence at $T_6 \simeq 250$. In fact, the cross-sections here vary slowly with the energy (IV-B), so that the choice is not critical.

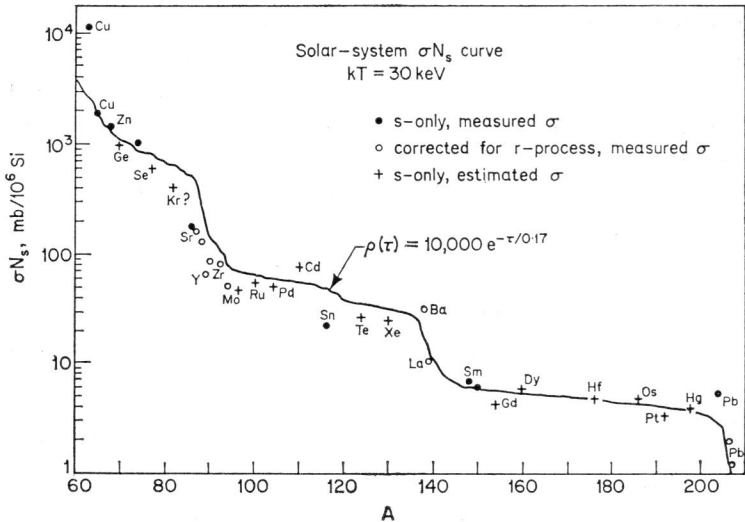

Fig. IV-6. $\sigma_{n,\gamma}N_s$ diagram of the $s$ elements. The points form a monotonically decreasing curve.

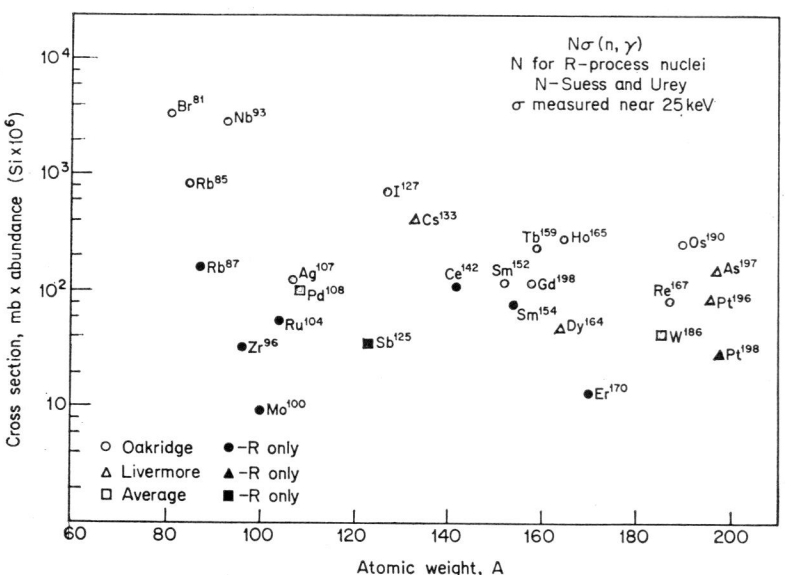

Fig. IV-7. $\sigma_{n,\gamma}N_r$ of the $r$ elements; no correlation.

Integration gives $\psi_A(\tau)$, where $\tau$ is the total irradiation, which can be converted into $\psi_A(n_C)$, where $n_C$ is the number of neutrons absorbed by an Fe atom. For each value of $\tau$, we can calculate $n_C$ by means of the equation

$$n_C = \sum_{A=56}^{209} (A - 56) n_A = \sum_{A=56}^{209} (A - 56) \frac{\psi_A}{(\sigma_A)} \qquad \text{IV-10}$$

At the beginning of the irradiation (at $\tau \simeq 0.1$), $n_C \simeq 30\tau$. Then $n_C/\tau$ increases rapidly until $\simeq 90$ (for $\tau = 1$) and then decreases slowly towards 60 ($\tau \simeq 2.5$). Fig. IV-8 illustrates the behaviour of $\psi_A = (\sigma_A N_A)(\tau)$ in terms of $n_C$.

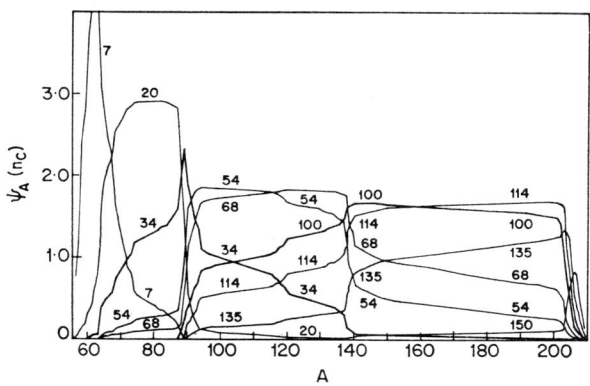

FIG. IV-8. Calculation of the distributions $\psi_A = \sigma_A N_A(n_C)$ after various intensities of neutron irradiation. On each curve, the number $n_C$ of neutrons absorbed per seed-nucleus (Clayton, 1962).

It is obvious that the curve in Fig. IV-8 does not resemble the curve IV-6 for any value of $n_C$ (or of $\tau$).

We deduce that the observed curve IV-6 is not the result of the irradiation of a unique target, but rather of quite a few irradiated targets with different $\tau$ (IV-C-6).

We define a fraction $\rho(\tau)$ which represents the fraction of the Fe nuclei subjected to an irradiation $\tau$, $\left( \int \rho(\tau) \, d\tau = 1 \right)$ or, parallelly, $g(n_C)$ the fraction of the Fe nuclei which have absorbed $n_C$ neutrons

$\left(\int g(n_C)\, dn_C = 1\right)$. Then we determine the functions $g(n_C)$ and $\rho(\tau)$ by means of the equation

$$\sigma_A N_A \text{ (observed)} = \int_0^\infty g(n_C)\psi(n_C)\, dn_C = \int \rho(\tau)\psi(\tau)\, d\tau \qquad \text{IV-11}$$
$$\simeq 4 \times 10^{-3}\psi(n_C = 3) + 10^{-3}\psi(n_C = 7)$$
$$+ 10^{-4}\psi(n_C = 35) + 10^{-5}\psi(n_C = 100)$$

or also:

$$\rho(\tau) \simeq 0\cdot 02 \exp\left(\frac{-\tau}{0\cdot 17}\right)$$

The determination is not unique, of course, but the solutions obtained have certain qualitative features which are probably real. The most striking fact is that at most 1% of the stellar iron nuclei has been irradiated by neutrons and, of these, at most $10^{-4}$ received large irradiations. Why? We will come back to it (Lecture V).

### IV–F: *The r elements*

Fig. IV–9 illustrates their abundance and Figs. IV–4 and IV–10 illustrate their mode of formation. Here we assume that $t_\beta \gg t_{n,\gamma}$: in the $Z$–$N$ plane, the horizontal path continues. As we go further from the bottom of the valley, the stability of the encountered nuclei decreases rapidly. The lifetime with respect to $(\gamma, n)$ disintegration decreases

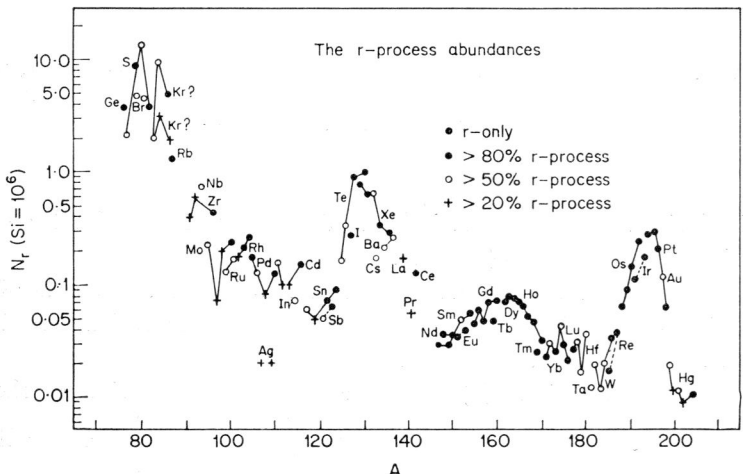

FIG. IV–9. The abundance of the *r* elements.

correspondingly (II–D). The path leads us to a "waiting" nucleus, $A$, characterized by $t_{n,\gamma} = t_{\gamma,n}$. There, we mark time between $A$ and $A + 1$ during time $t_\beta$. Then we go up again and continue. It is obvious that the path keeps well to the right of the stability valley.

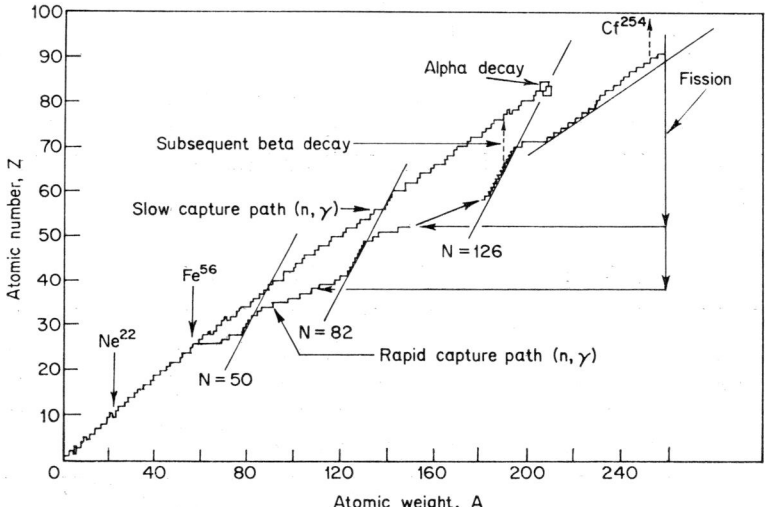

Fig. IV–10. The $r$ path. The fission does not seem to play an important role

When the number $N = 50$ is reached, the path returns towards the stability valley. The 51st neutron is very little bound (its $t_{\gamma,n}$ is very short). The isotopes $N = 50$ are hence waiting isotopes, and we get closer to the valley until the time $(\gamma, n)$ is long enough to allow the addition of the 51st neutron.

The same phenomenon occurs at $N = 82$ and $N = 126$. As for the $s$ case, the population density of the elements along the $r$ path is proportional to the flow velocity. Since here $t_\beta \gg t_{n,\gamma}$, it is obvious that the isotopes will be concentrated on the steps $N = 50, 82, 126$, where it is $t_\beta$ which governs the flow. In general, the $t_\beta$ increase as we approach the valley, so that the maximum concentration is at the top of the steps.

When irradiation is finished, all the elements move by $\beta^-$ disintegration (hence vertically in Fig. IV–10) towards the stability valley.

The elements concentrated at the top of the steps form three abundance peaks. Since the $A$ of the step elements is *smaller* than the $A$ of the stable elements with the same number of neutrons, the peaks produced in this way are to the *left* of the peaks of the $s$ elements. In the first peak we find Se, Kr, in the second Te, T, Xe, and in the third Or, Ir and Pt.

Finally, we must attribute to the $r$ process the existence of the transuranian elements (the $s$ process was restricted to the elements with $A \leqslant 209$). These elements can fission and contribute to the abundance distribution. This contribution does not seem very large.

A quantitative study of the $r$ process is very difficult since it requires the knowledge of the nuclear systematics of quite unstable elements. There have not yet been many results. It has been shown that the ambient conditions must be $T_6 \simeq 1000$ and $n_n = 10^{20 \text{ to } 25}/\text{cm}^3$; and $\Delta t \simeq 1$ second.

### IV–G: *Location of the s and r mechanisms*

It is fairly clear that we must assign to the irradiations of neutrons accompanying the *stable* periods of stellar evolution the task of producing $s$ elements. Choosing temperatures ranging from $T_6 = 100$ to $1000$ does not affect the monotonicity if the $\sigma_s N_s$ curve and the times $t_{n,\gamma}$ are much longer than most of the times $t_\beta$.

The question of the $r$ elements is not as clear. Two types of locations occur: the stellar flashes (those of carbon and of hydrogen are still hypothetical) and the supernovae, of which we know very little. Fig. III–9 presents the evolution of the density and of the temperature according to a preliminary model. It seems that we find here appropriate conditions.

### IV–H: *The p elements*

In Fig. IV–3 they are the elements to the left of the valley. By comparison with the $s + r$ elements of same $A$, they are always very rare ($\simeq 1\%$). Little is known about their origin.

They can be produced by $(\gamma, n)$ reactions on $s$ elements, for instance in the flux of gamma rays resulting from a stellar explosion (supernovae) (Fig. III–9). In the absence of detailed models of supernovae, it is impossible to calculate the abundances.

They can also be produced by $(p, \gamma)$ protons. Because of the Coulomb barrier, we must then have very high temperatures. At these temperatures, the protons have long since disappeared. We must call upon a mechanism of hydrogen injection in a supergiant. And even then there are difficulties . . .

The nuclear reactions at the surface of stars could also be invoked.

And also the action of the positrons of the stellar gas in Bose–Einstein equilibrium on the $s$ elements which are already produced. The existence of these positrons has been discussed in the preceding lesson. In the table we see their density for some $T$-$\rho$ ($n^+/\text{cm}^3$).

| $T_6$ | $\rho = 10^6$ gm/cm$^6$ | $= 10^5$ gm/cm$^6$ |
|---|---|---|
| 1000 | $n^+/\text{cm}^6 = 4 \times 10^{26}$ | $10^{25}$ |
| 2000 | $4 \times 10^{28}$ | $2 \times 10^{28}$ |

The "tunnel" process for a reaction $(A + e^+ \to A + \nu)$ is much easier than for a reaction $(A + p \to (A + 1) + \gamma)$ of same energy $(P \propto \exp(-2\pi Z_1 Z_2 e^2/\hbar v))$.

We illustrate the process with the case of $_{68}\text{Er}^{164}$. The corresponding element, $_{66}\text{Dy}^{164}$, is formed, say, at the time of the helium-burning phase ($T_6 \simeq 200$). Then, at the time of the $\text{O}^{16}$-burning phase, the $_{66}\text{Dy}^{164}$ captures an $e^+$. At $T_6 = 1500$, the capture takes $\simeq 30$ years. We form thus $_{67}\text{Ho}^{164}$, which $\beta^-$ disintegrates part of the time towards $_{68}\text{Er}^{164}$.

All these $p$-forming processes have their merits. However, neither of them explains the observed regularity of the $p$ abundances in terms of $A$ (Fig. I-9). Would there be an even more important mechanism?

## Bibliography

A good start:

the article by FELD in "Experimental Nuclear Physics", edited by Segrè (Wiley, 1961)

The experimental results of interest in neutron astrophysics:

R. L. MACKLIN, J. H. GIBBONS, *Rev. Mod. Phys.* **37**, 166, 1965
H. REEVES, "Stellar Neutron Sources", *Ap. J.* **146**, 447, 1966

On the ($s$ and $r$) processes (also "Nuclear Astrophysics" by A. G. W. CAMERON):

D. D. CLAYTON, W. A. FOWLER, T. E. HULL, B. A. ZIMMERMAN, "Neutron Capture Chains in Heavy Element Synthesis", *Annals of Physics*, **12**, 331, 1961
M. P. LACROUTE, M. M. TIMORES, "Abondance des isotopes synthétisés par un processus de capture neutronique", *J. de Phys.* **24**, 273, 1963
P. A. SEEGER, D. D. CLAYTON, W. A. FOWLER, "Nucleosynthesis of Heavy Elements by Neutron Capture", 1965 *Ap. J.* Suppl., in press
D. D. CLAYTON, "Implications of the Solar-System Abundances near Atomic Weight 90", *Journal of Geophysical Research*, **69**, 5081, 1964

On the $p$ process:

G. MALKIEL, *A.J. U.R.S.S.* **7**, 207, 1963
H. TSUJI, *Progr. Theor. Phys.* **27**, 608, 1962

F. KAMENETSKII, *A.J. U.R.S.S.* **5**, 66, 1961

H. REEVES, P. STEWART, "Positron Capture as a possible Source of the $p$ Elements", *Ap. J.* **141**, 1432, 1965

Some observations:

J. L. GREENSTEIN, G. WALLERSTEIN, *Ap. J.* **139**, 1163, 1964

# V

## The evolution of the galaxy; stellar statistics and cosmochronology

WE have distinguished between two sorts of H–R diagrams: those which refer to determined clusters (hence to stars of similar age and chemical composition), and those of the set of stars called the "field stars" of the solar neighbourhood, which represent the branch of the galaxy in which the sun is located, hence in all likelihood of different ages and chemical compositions. We have discussed the first sort (Lecture II), we will now consider the second. A programme for their study is presented in the next section; in theory it is simple, but it cannot be applied in practice. Starting from this programme, more realistic methods are considered and their results are discussed.

### V–A: *Definitions and theoretical programme*

Consider a point density function $\phi(L, T_e)$ which represents the actual appearance ($t = 0$) of the H–R diagram of the set of field stars.

Let there be a function $\xi(M, t_0)$ representing the rate of creation of stars of mass $M$ at time $-t_0$ in the past; hence the rate of injection of the H–R points in the H–R plane.

Let there be a function $S(M, c, -t + t_0)$ describing the path of a star of mass $M$, of chemical composition $c = \sum n_i$, born at time $t_0$, as a function of its age. Assuming a good mixing of the galactic gas, we have $c = c(t_0)$ (uniquely).

If $\xi$ and $S$ were known for all $M$, $c$ and $t_0$, and if $c(t_0)$ were known, we could calculate the function $\phi(L, T_e)$ and compare it with observations.

Let $\Delta(M)$ be the fraction of mass returned to the galaxy, quietly or violently, by the star of mass $M$ (perhaps $\Delta M = \Delta M(M, c)$?).

Let $\Delta n_i(M, c)$ be the amount of $i$ isotopes returned by the star $M, c$; $\Delta M = \sum M_i \Delta n_i$.

Knowledge of $\Delta(M)$ and of $\Delta n_i$ would allow us to calculate $c(t_0)$ and to study the evolution of the galaxy since its origin. From this, we would obtain, for instance, the variation in total luminosity, in colour, in chemical abundances, in terms of time.

We would then have to consider the variations of these quantities in the interior of the galaxy, the local inhomogeneities and the mixing velocities. Then we would obtain a theory of galactic evolution analogous to that of stellar evolution. This is the next step in the study of the cosmos.

V-B: *Simple model; the density function of the main sequence*

Of all the functions defined above, the least known is $\xi(M, t_0)$. The mechanism which leads a mass of gas to isolate itself from the interstellar gas and to develop individually remains more or less unknown. We will hence try to determine $\xi$ from observations, or rather, we are going to take a set of $\xi$ and try to reconstitute $\phi$.

The function $S$ has been discussed in Lecture II. Comparatively well established until the red-giant phase, the path escapes us during the last moments of the life of a star (at least for large stars), but we expect to find it back again in the white-dwarfs region.

We will use a simplified version of $S$. The star goes to the main sequence in a time zero. It remains there for a time $t = t_\mathrm{H}$ (time which is necessary for the fusion of 10% of the initial hydrogen). Then it leaves the main sequence, and becomes instantaneously a white dwarf. Consequently, we will concentrate on the function $\phi$ of the main sequence. The observed function presents a sudden change in average slope ("a knee").

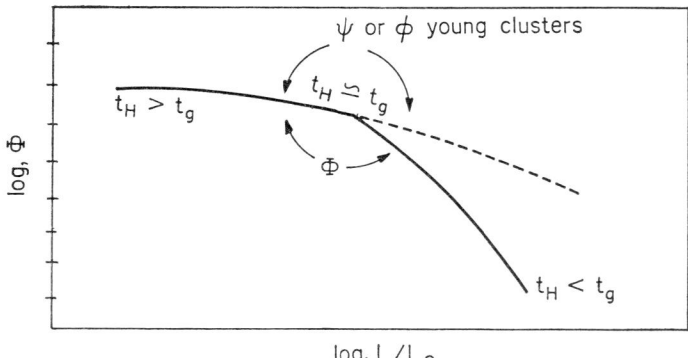

Now this break occurs at a point where the time $t_H$ of the corresponding star (see Lecture II) is more or less equal to the age of the galaxy, $t_g$. Moreover, the function $\phi$ of the very young clusters has a shape very similar to the left-hand part of the diagram ($t_H > t_g$), but does *not* have a knee.

## V–C: *Rate of star creation*

The simplest interpretation: (1) the rate of creation is independent of time and, (2) $\xi = \xi(M)$ can be obtained from the observed $\phi$ simply by the following relations:

for $L$ such that $t_H > t_g$: $\xi = +\phi/t_g$ (uniform rate of creation since the beginning of the galaxy);

for $L$ such that $t_H < t_g$: $\xi = \phi/t_H$ (uniform rate of creation since $t_H$; the preceding ones are gone!).

With this choice of $\xi$, we find the following remarkable result: The function $\psi = \int_{t_g}^{0} \xi(M)\, dt = \xi(M) t_g$, which represents the sum of all the $M$ stars which ever existed, (1) does not have a knee and (2) now resembles *everywhere* the function of the very young clusters. This agreement is certainly not accidental, since at the greatest luminosities the curves $\phi$ and $\psi$ differ by a factor of more than one thousand (diagram).

From this we can also calculate the ratio of the number of stars which left the M.S. to the number of stars which remained there. We obtain a number of the order of 10%. This 10% also represents rather well the ratio of the number of white dwarfs to the number of stars on the M.S. The hypothesis of time independence is thus not a very bad one.

To what extent is it valid? After all it can be objected that star formation occurs at the expense of the galactic gas, the total mass of this gas must have decreased with time and, consequently, the rate of creation must also have decreased.

Now, in the solar neighbourhood, the gaseous fraction of the galactic mass is of only a few per cent.

Moreover, we have several reasons to believe that about 20% of the galactic mass condensed very early into stars (in less than $10^9$ years), at the very beginning of the galaxy. These stars (extreme Pop. II) now form the globular clusters of the galactic halo. How can these statements be reconciled with the results of the preceding paragraph?

### V–D: *Evolution of the galactic gas*

In order to study it, we will merely assume (1) that $\xi$ is a separable function, $\xi(M, t_0) = \xi(M)f(t_0)$; (2) that $\xi(M)$ is given for young clusters; (3) that $f(t_0) \propto \rho_g^n(t_0)$ where $\rho_g(t)$ is the density of the galactic gas at time $t$ ($n$ is an exponent which must be determined) and we will follow the evolution of the galactic gas $M_g(t)$, taking into account the condensation into stars and the restitution. To do this, we must determine $\Delta M(M)$.

Since the observed white dwarfs have average masses of $\simeq 0.7 M_\odot$, we simply choose $\Delta M = (M - 0.7 M_\odot)$ and we write

$$m_g(t) = m_g(t_g) - m_T(t) + m_E(t) \qquad \text{V-1}$$

$$m_T(t) = \int M \xi(M)\, dM \int_{t_g}^{t} f(t)\, dt \qquad \text{V-2}$$

where $M_T(t)$ is the total mass of the stars formed up to time $t$, and

$$m_E(t) = \int \xi(M)(M - 0.7 M_\odot) \int_{t_g}^{t-t_H(M)} f(t)\, dt\, dM \qquad \text{V-3}$$

where $M_E(t)$ is the total mass ejected by the stars in evolution.

Salpeter chose $n = 1$ and showed that then the galactic gas decreased more or less exponentially with a half-life of roughly $10^9$ years. The initial stellar activity was very rapid, less than $3 \times 10^8$ years were required to make 20% of the stars (our extreme Pop. II stars?).

Notwithstanding this large variation of $f(t)$, the calculated $\phi$ remains rather close to the observed $\phi$ (within the difficulties of observation). It is mostly the young stars which influence this comparison, and the long period during which $\rho_g(t)$ decreases shows that the rate of creation changed very little since the birth of these stars.

Schmidt takes $n$ as a parameter, and tries to explain other observed phenomena at the same time. He concludes that $n$ must vary with $M$, hence that the function $\xi$ is not exactly separable ($\xi = \xi(M)f(t) + \xi'(M, t)$). Because of observational uncertainties, it is difficult to determine the importance of the term $\xi'(M, t)$.

### V–E: *Nucleosynthesis of the groups of elements*

Here we group together the products of the various nucleosynthetic mechanisms: helium, representing roughly 20% of the galactic mass;

CNO, $(10^{-2})$; the elements from neon to sulphur, $(10^{-3})$; the iron peak, $(5 \times 10^{-4})$; the heavy elements, $(\simeq 5 \times 10^{-6})$; the very heavy elements, $(\simeq 10^{-8})$. In order to study theoretically the evolution of these abundances, we must know the fraction of the stellar mass contained in the layers containing each of these products. This has been done rather well for the hydrogen and helium stages (Fig. V-1), but no further. We know neither the final distribution nor the changes the violence of the ejection can make in it. Notwithstanding these difficulties, some results are becoming clear.

The increase with age of elements heavier than helium (especially iron) reaches a factor 10. On the other hand, the helium abundance does not seem to present such an increase. Did the initial galactic gas contain helium? Where would this helium come from? From earlier galaxy generations? From the "big bang"?

Hence we now have the following results: less than 20% (in mass) of

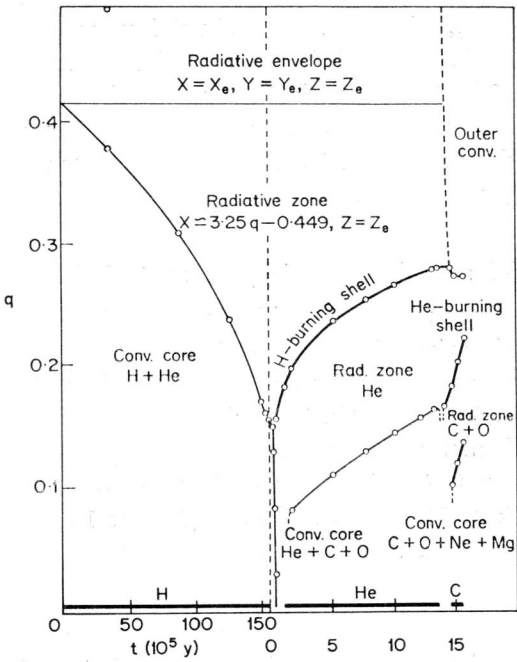

FIG. V-1. The evolution of the structure and of the chemical composition of a star of $15 \cdot 6 M_\odot$ (Hayashi, 1962). The curves show the locus of the zones of nuclear burning and the boundaries of the convective regions. ($q = M_r/M$) is the fraction of stellar mass contained between the centre and the point under consideration $(r)$.

the galactic hydrogen went at least as far as the stage of hydrogen burning. Of the helium produced in this way, less than 5% went as far as the stage of helium burning. Of the carbon and the oxygen produced in this way, less than 10% is transformed (in two stages) into the elements from neon to sulphur. Of these elements, more than 50% produced the elements of the iron peak. Only 1% of the iron peak was affected by neutrons and less than $10^{-4}$ of these seed-nuclei received more than 50 neutrons. (These numbers obviously only apply to stellar matter in "circulation". The internal chemical distribution of white dwarfs is forever beyond reach of our observation.)

The first three fractions are small, but the following one is rather more important; this difference is probably connected with the difference in the synthesizing processes (the dominant factor goes from $Z$ to $\Delta M$). The smallness of the last factors can be explained in the following manner: the maximal mass of heavy elements ($A > 62$) ejected from a given star is more or less the mass of the iron nuclei in the original stellar gas ($\sim 10^{-3}$), whereas the maximal iron mass contains all the mass of the zone of the equilibrium process, less the mass of the remnant white dwarf ($\simeq 1 M_\odot$). If we make the reasonable hypothesis that the $s$ and $r$ processes are nearly always followed by the $e$ process, it suffices that the ejected iron mass be 10% of the total mass to explain the 1% of the preceding paragraph.

### V–F: *Nucleosynthesis of the unstable isotopes*

The existence in nature of unstable elements gives us a very powerful tool for evaluating times in Astrophysics.

If we make exception of the elements formed in all probability by cosmic radiation (such as $C^{14}$), there do not seem to exist, in the solar system, any elements with mean life $\ll 10^9$ years, but there exist several elements with $t \geqslant 10^9$ years, such as $U^{235}$ (half-life $= 1 \cdot 03 \times 10^9$ years), $U^{238}$ ($6 \cdot 54 \times 10^9$ years), and $Th^{232}$ ($13 \cdot 9 \times 10^9$ years). This simple fact gives a lower limit to the age of the elements.

### V–G: *The age of the r elements and the age of the galaxy*

When the creation mechanisms have stopped, the abundance of an unstable isotope with mean life $\tau_i$ varies as

$$X_i(t) = X_i(0) \exp\left(-\frac{t}{\tau_i}\right) \qquad \text{V-4}$$

V–G–1) If we assume that the element $i$ was synthesized very suddenly (in $\Delta t \ll \tau_i$; *instantaneous synthesis*) the equation V–3 gives a good description of the evolution of its abundance. But how shall we determine $X_i(0)$? We will hence rather analyse the ratio of two isotopes, for example $U^{235}$ and $U^{238}$.

$$\frac{X_{235}(t)}{X_{238}(t)} = 0\cdot 0072 = \frac{X_{235}(0)}{X_{238}(0)} \exp\left(-\frac{t}{\tau_{235}} + \frac{t}{\tau_{238}}\right) \qquad \text{V–5}$$

The value $0\cdot 0072$ is the actual value for natural uranium. The ratio $X_{235}(0)/X_{238}(0)$ is the rate of relative synthesis in the $r$ process (Lecture IV). This rate is not very well known, but must have a value close to 1. The calculated age is then $6 \times 10^9$ yrs. The same calculation performed with $U^{238}$ and $Th^{232}$ gives more or less the same result. The age determined in this way would be, for instance, the time gone by since the explosion of a unique hypothetical supernova from which all our heavy elements would originate.

V–G–2) In reality, everything rather suggests a gradual enriching of the galactic gas. We will call upon a *continuous synthesis* which we will assume, for the time being, to be *constant* (see by analogy, Section V–B). We have:

$$\frac{dX_{235}}{dt} = \lambda_{235} - \frac{X_{235}}{\tau_{235}} \qquad \text{V–6}$$

where $\lambda_{235}$ is the rate of creation of $U^{235}$ by the $r$ process, assumed to be constant from $t = 0$ to $t_i$ (the time when, since the protosolar gas has started to condense, it has stopped receiving contributions from the contemporary supernovae).

At $t_i$ the ratio of the uraniums is given by

$$\frac{X_{235}(t_i)}{X_{238}(t_i)} = \frac{\tau_{235}\lambda_{235}\{1 - \exp(-t_i/\tau_{235})\}}{\tau_{238}\lambda_{238}\{1 - \exp(-t_i/\tau_{238})\}} \qquad \text{V–7}$$

Then from $t_i$ to $t$ (now), we have

$$\frac{X_{235}(t)}{X_{238}(t)} = \frac{X_{235}(t_i)}{X_{238}(t_i)} \frac{\exp\{+(t_i - t)/\tau_{235}\}}{\exp\{+(t_i - t)/\tau_{238}\}} \qquad \text{V–8}$$

The calculation leaves one parameter free, let us pick $|(t - t_i)|$: the time since the end of the nucleosynthesis of the elements of the solar system. The minimal value is the age of the sun ($\simeq 4\cdot 5 \times 10^9$). Giving to $(t - t_i)$ the values $4\cdot 5$, $5\cdot 5$, $6\cdot 6$, we find for $t$ (the time since the beginning of the nucleosynthesis of the $r$ elements) the values 18, 8·5,

6·6. The case 6·6 corresponds to the instantaneous synthesis with the hypothesis $X_{235}(0)/X_{238}(0) = 1\cdot 6$ (cf. Fig. V–2).

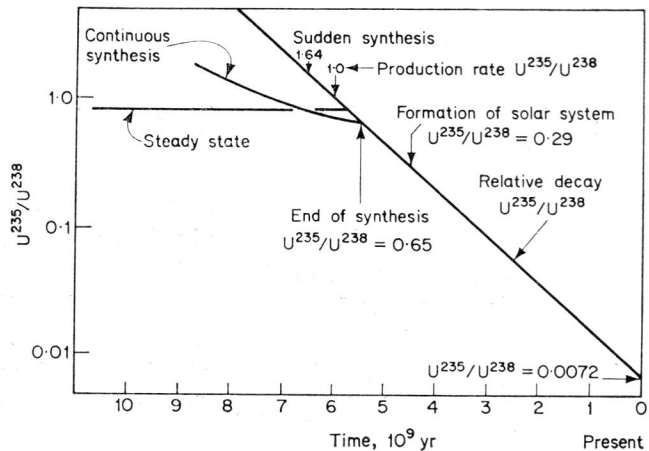

FIG. V–2. The ratio $U^{235}/U^{238}$ as a function of time (Burbidge, 1957). The various curves in the left half of the diagram correspond to various hypotheses on the variation with time of the rate of synthesis of the $r$ elements.

V–G–3) Always in analogy with Sections V–B and V–C, we now introduce the hypothesis of a *continuous synthesis* but whose variation with time is *connected with the (exponential) evolution of the galactic activity*. Here the time $t = 0$ will be the origin of the galaxy. We must first wait for a time $t_r$ before the $r$ elements start appearing; this is the lifetime of the stars which can produce these elements. In all likelihood, $t_r \leqslant 10^9$ years. From $t_r$ to $t_i$: increasing synthesis, then from $t_i$ to $t$: exponential disintegration. This model allows us to calculate $|t_r - t_i| \simeq 7 \times 10^9$ years. Whence we can estimate the age of the galaxy: $T_g \simeq (t_r + 12) \times 10^9$ years $\simeq 1\cdot 2 \times 10^{10}$ years. This age is similar to the ages of the oldest star clusters (whose age is determined from the H–R paths).

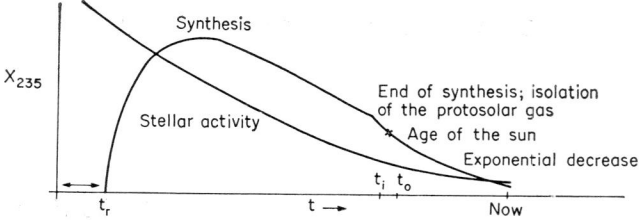

The fractional abundance of uranium in the gas which formed the solar system.

## Bibliography

Stellar statistics and stellar evolution:

R. J. TRUMPLER, H. F. WEAVER, "Statistical Astronomy", Berkeley University of California Press (1953)

E. E. SALPETER, "The Luminosity Function and Stellar Evolution", *Ap. J.* **121**, 161, 1955
 "Statistics of Stellar Evolution", Académie Pontificale des Sciences (1958)
 "The Rate of Star Formation in the Galaxy", *Ap. J.* **129**, 608, 1959
 Symposium de Solvay (1964) (to appear)

M. SCHMIDT, "The Rate of Star Formation", *Ap. J.* **129**, 243, 1959
 *Ap. J.* **137**, 758, 1963

A. SANDAGE, *Ap. J.* **125**, 422, 1957

S. VAN DEN BERGH, *Ap. J.* **125**, 445, 1957
 *Z. of Ap.* **43**, 236, 1957

M. SCHMIDT, "The Evolution of the Sun's Neighbourhood", Symposium on Stellar Evolution, La Plata Observatory, Argentina

D. N. LIMBER, "The Universality of the Initial Luminosity Function", *Ap. J.* **131**, 168, 1960

Cosmochronology:

W. A. FOWLER, F. HOYLE, "Nuclear Cosmochronology", *Ann. Phys. N.Y.* **10**, 280, 1960
 "On the Abundances of U and Th in the Solar-System Material" (1963). In "Isotopic and Cosmic Chemistry", North-Holland Publishing Co., Amsterdam

The helium question:

R. J. TAYLOR, H. HOYLE, "The Mystery of the Cosmic Helium Abundance", *Nature*, **203**, 1108, 1964

Helium in the evolutionary and cosmochronological context:

J. W. TRURAN, J. C. HANSEN, A. G. W. CAMERON, "The Helium Content of the Galaxy", in "Stellar Evolution", Plenum Press, New York